Differential Equation Models in Applied Mathematics: Theoretical and Numerical Challenges

Differential Equation Models in Applied Mathematics: Theoretical and Numerical Challenges

Editor

Fasma Diele

MDPI • Basel • Beijing • Wuhan • Barcelona • Belgrade • Manchester • Tokyo • Cluj • Tianjin

Editor
Fasma Diele
Istituto per le Applicazioni del
Calcolo M. Picone, CNR
Italy

Editorial Office
MDPI
St. Alban-Anlage 66
4052 Basel, Switzerland

This is a reprint of articles from the Special Issue published online in the open access journal *Mathematics* (ISSN 2227-7390) (available at: https://www.mdpi.com/journal/mathematics/special_issues/Differential_Equation_Models).

For citation purposes, cite each article independently as indicated on the article page online and as indicated below:

LastName, A.A.; LastName, B.B.; LastName, C.C. Article Title. *Journal Name* **Year**, *Volume Number*, Page Range.

ISBN 978-3-0365-3010-9 (Hbk)
ISBN 978-3-0365-3011-6 (PDF)

© 2022 by the authors. Articles in this book are Open Access and distributed under the Creative Commons Attribution (CC BY) license, which allows users to download, copy and build upon published articles, as long as the author and publisher are properly credited, which ensures maximum dissemination and a wider impact of our publications.
The book as a whole is distributed by MDPI under the terms and conditions of the Creative Commons license CC BY-NC-ND.

Contents

About the Editor vii

Fasma Diele
Differential Equation Models in Applied Mathematics: Theoretical and Numerical Challenges
Reprinted from: *Mathematics* **2022**, *10*, 249, doi:10.3390/math10020249 1

Oana Brandibur, Roberto Garrappa and Eva Kaslik
Stability of Systems of Fractional-Order Differential Equations with Caputo Derivatives
Reprinted from: *Mathematics* **2021**, *9*, 914, doi:10.3390/math9080914 5

Deborah Lacitignola
Handling Hysteresis in a Referral Marketing Campaign with Self-Information. Hints from Epidemics
Reprinted from: *Mathematics* **2021**, *9*, 680, doi:10.3390/math9060680 25

Elishan Christian Braun, Gabriella Bretti and Roberto Natalini
Mass-Preserving Approximation of a Chemotaxis Multi-Domain Transmission Model for Microfluidic Chips
Reprinted from: *Mathematics* **2021**, *9*, 688, doi:10.3390/math9060688 43

Gabriele Vissio and Antonello Provenzale
On-Off Intermittency in a Three-Species Food Chain
Reprinted from: *Mathematics* **2021**, *9*, 1641, doi:10.3390/math9141641 77

Alyona Zamyshlyaeva and Aleksandr Lut
Inverse Problem for the Sobolev Type Equation of Higher Order
Reprinted from: *Mathematics* **2021**, *9*, 1647, doi:10.3390/math9141647 89

Ahmed AlGhamdi, Omar Bazighifan and Rami Ahmad El-Nabulsi
Important Criteria for Asymptotic Properties of Nonlinear Differential Equations
Reprinted from: *Mathematics* **2021**, *9*, 1659, doi:10.3390/math9141659 103

Francesca Mazzia and Giuseppina Settanni
BVPs Codes for Solving Optimal Control Problems
Reprinted from: *Mathematics* **2021**, *9*, 2618, doi:10.3390/math9202618 113

About the Editor

Fasma Diele (MSc in Mathematics, University of Bari, Italy, 1991) is a senior researcher in the 'Environmental Mathematics' Project Area of IAC-CNR. She is the author of more than 50 papers, mainly in the field of Numerical Analysis and Applied Mathematics, with more than 400 citations and an H-index of 11. She has been involved in environmental modelling and numerical research activities within several EU funded projects (BIO_SOS, H2020 ECOPOTENTIAL, eLTER-Plus, and CHOECO). She has held an Italian national scientific qualification for Associate Professor in Numerical Analysis since 2012. She is a member of the Faculty Board of PhD in Mathematics, University of Bari (until 2017), Editor for 'Abstract and Applied Analysis' by Hindawi (censed by WOS) and other minor journals, an active member of the GNCS group of INDAM (Istituto Nazionale di Alta Matematica), and an elected member of Giunta of MSE group of UMI (Unione Matematica Italiana). She also serves as a scientific evaluator for ANVUR-VQR 2015-1019 (Valutazione della Qualità della Ricerca) and a research project evaluator for the Italian Ministero dello Sviluppo Economico (MiSE), for the Italian Ministero dell'Istruzione dell'Università e della Ricerca (MIUR), and for National Research Fund for Scientific & Technological Development (FON-DECYT), Chile, in the Mathematics area. Additionally, she is a member of European Women in Mathematics (EWM).

Editorial

Differential Equation Models in Applied Mathematics: Theoretical and Numerical Challenges

Fasma Diele

Istituto per Applicazioni del Calcolo 'M.Picone', CNR, Via Amendola 122/D, 70126 Bari, Italy; fasma.diele@cnr.it

1. Motivations for the Special Issue

The articles published in the Special Issue "Differential Equation Models in Applied Mathematics: Theoretical and Numerical Challenges" of the MDPI *Mathematics* journal are here collected. The Special Issue intended to highlight old and new challenges in the formulation, solution, understanding, and interpretation of models of differential equations (DEs) in different real world applications. Indeed, models of differential equations can describe complex mechanisms arising in a wide range of applications in many different sectors as ecology, health, biology, economics, and finance. Differential modelling and difference equations are tools to understand the dynamics, do forecasting and scenario analysis; in addition, they allow for the detection of optimal solutions according to selected criteria.

The technical topics covered in the seven articles published in this book include: asymptotic properties of high order nonlinear DEs [1,2], analysis of backward bifurcation [3], stability analysis of fractional-order differential systems [4]. Models oriented to real applications consider the chemotactic between cell species [5], the mechanism of on-off intermittency in food chain models [6] and the occurrence of hysteresis in marketing [3]. Numerical aspects deal with the preservation of mass and positivity [5] and the efficient solution of Boundary Value Problems (BVPs) for optimal control problems [7].

In the following, I summarize the main content of novelty of this book distinguishing among contributes that concerns:

- Theoretical challenges of DEs [1,2,4];
- Numerical challenges of DEs [5,7];
- Real-word applications of DEs [3,6].

1.1. Theoretical Challenges of DEs

In articles [1,2] the focus is on high-order differential equations. In [1], new oscillation theorems for fourth-order differential equations are established using the Riccati and the integral averaging techniques. The article [2] investigates the inverse problem for a non homogeneous, higher-order Sobolev type equation with assigned Cauchy and overdetermined conditions. By using the theory of bounded polynomial operator pencils, the problem is initially reduced into two regular and singular aggregates and then it is restored using the method of successive approximations. A theorem on the solvability of the original problem represents the main theoretical contribute of this paper to the literature on this subject.

In the review article [4] systems of fractional-order DEs with Caputo derivative are presented. Due to the dependence on the order of the fractional derivatives, the linear stability analysis leads to properties fundamentally different from those of classical DEs: unlike systems of integer order, coefficients of the systems are not sufficient to describe stability properties of solutions. By reviewing the asymptotic analysis of the Mittag–Leffler function and of its derivatives and by examining systems with some specific structures, this paper intends to contribute to the research on the stability analysis of multi-order higher dimensional systems that represents nowadays an important theoretical challenge for fractional-order DEs.

1.2. Numerical Challenges of DEs

The transmission model for microfluidic chips presented in the featured paper [5] involves doubly parabolic DEs in 2D spatial domains connected with either a doubly parabolic or a hyperbolic-parabolic DEs in 1D domains. The important contribute of this paper is the development of novel positive numerical conditions and numerical methods, based on finite difference schemes, assuring the mass-preservation at the external boundaries and the interfaces between domains of different sizes. It has to be underlined that is the first numerical work where this new technique of switching the size of the domains and type of partial differential equations, i.e., parabolic vs. hyperbolic, is introduced in the literature.

In paper [7], Hamiltonian boundary valued DEs deriving from the applications of the indirect method, based on Pontryagin's conditions, to optimal control problems are considered. The main contribute of this paper is to show how to properly choose and use codes on popular scientific platforms (Fortran, Matlab, R), for solving some specific challenging optimal control problems. This paper gives important indications useful to choose an initial mesh, to handle the input parameters or to use of a continuation technique for nonlinear problems to achieve accurate solutions via a bvp (boundary value problem) solver.

1.3. Real-Word Applications of DEs

In the featured paper [6], the power of DEs as leading mathematical tools for describing ecosystem dynamics is illustrated. In particular, some preliminary steps towards a conceptual description of population outbursts grounded into an environment-driven mechanism are described. The focus is on a three-species food chain represented by the Hastings–Powell model: by stochastically perturbing the value of some parameters, the authors show the emergence of on–off intermittency, i.e., an irregular alternation between stable phases and sudden bursts in population size. The strength of this paper lies in representing the first evidence of the possibility of on–off intermittent behavior in a food chains model.

The original point of view adopted in the paper [3] illustrates the ability of DEs in modelling and anlyzing dynamical scenarios in different real-word applications. Here the expert approach of the author adopts the analogy between the key mechanism of contagion for both the spread of an epidemic and for a referral marketing defined as viral because of it involves a person-to-person transmission. The theoretical concept of backward bifurcation, to avoid in the epidemic context, in the viral marketing could strengthen the campaign's chances of survival. However, the paper points out the possible introduction of a risk factor in the bistability range where, according to the chosen initial conditions, hysteresis-type behaviors can emerge.

2. Conclusions

I hope that this collection will be useful for those working in the area of modelling real-word applications through differential equations and those who care about an accurate numerical approximation of their solutions. The reading is also addressed to ones who are willing to become familiar with differential equations which, due to their predictive abilities, represent the main mathematical tool for making scenario analysis of our changing world [8].

Funding: This research received no external funding

Acknowledgments: As the guest editor, I want to thank all the authors for contributing to the Special Issue with their interesting and valuable articles collected in this book. I would also to thank all referees for their thorough and timely reports on the submitted works. Finally, it is my pleasure also to thank, in person of Grace Du, all the editorial staff of the journal Mathematics for the pleasant cooperation, during the preparation of the Special Issue and during the preparation of this book.

Conflicts of Interest: The author declares no conflict of interest.

References

1. AlGhamdi, A.; Bazighifan, O.; El-Nabulsi, R.A. Important criteria for asymptotic properties of nonlinear differential equations. *Mathematics* **2021**, *9*, 1659. [CrossRef]
2. Zamyshlyaeva, A.; Lut, A. Inverse Problem for the Sobolev Type Equation of Higher Order. *Mathematics* **2021**, *9*, 1647. [CrossRef]
3. Lacitignola, D. Handling Hysteresis in a Referral Marketing Campaign with Self-Information. Hints from Epidemics. *Mathematics* **2021**, *9*, 680. [CrossRef]
4. Brandibur, O.; Garrappa, R.; Kaslik, E. Stability of Systems of Fractional-Order Differential Equations with Caputo Derivatives. *Mathematics* **2021**, *9*, 914. [CrossRef]
5. Braun, E.C.; Bretti, G.; Natalini, R. Mass-preserving approximation of a chemotaxis multi-domain transmission model for microfluidic chips. *Mathematics* **2021**, *9*, 688. [CrossRef]
6. Vissio, G.; Provenzale, A. On-Off Intermittency in a Three-Species Food Chain. *Mathematics* **2021**, *9*, 1641. [CrossRef]
7. Mazzia, F.; Settanni, G. BVPs Codes for Solving Optimal Control Problems. *Mathematics* **2021**, *9*, 2618. [CrossRef]
8. Cuddington, K.; Fortin, M.J.; Gerber, L.; Hastings, A.; Liebhold, A.; O'connor, M.; Ray, C. Process-based models are required to manage ecological systems in a changing world. *Ecosphere* **2013**, *4*, 1–12. [CrossRef]

Review

Stability of Systems of Fractional-Order Differential Equations with Caputo Derivatives

Oana Brandibur [1], Roberto Garrappa [2,3] and Eva Kaslik [1,*]

[1] Department of Mathematics and Computer Science, West University of Timişoara, 300223 Timisoara, Romania; oana.brandibur@e-uvt.ro
[2] Department of Mathematics, University of Bari, Via E. Orabona 4, 70126 Bari, Italy; roberto.garrappa@uniba.it
[3] Member of the INdAM Research Group GNCS, Istituto Nazionale di Alta Matematica "Francesco Severi", Piazzale Aldo Moro 5, 00185 Rome, Italy
* Correspondence: eva.kaslik@e-uvt.ro

Abstract: Systems of fractional-order differential equations present stability properties which differ in a substantial way from those of systems of integer order. In this paper, a detailed analysis of the stability of linear systems of fractional differential equations with Caputo derivative is proposed. Starting from the well-known Matignon's results on stability of single-order systems, for which a different proof is provided together with a clarification of a limit case, the investigation is moved towards multi-order systems as well. Due to the key role of the Mittag–Leffler function played in representing the solution of linear systems of FDEs, a detailed analysis of the asymptotic behavior of this function and of its derivatives is also proposed. Some numerical experiments are presented to illustrate the main results.

Keywords: fractional differential equations; stability; linear systems; multi-order systems; Mittag–Leffler function

Citation: Brandibur, O.; Garrappa, R.; Kaslik, E. Stability of Systems of Fractional-Order Differential Equations with Caputo Derivatives. *Mathematics* **2021**, *9*, 914. https://doi.org/10.3390/math9080914

Academic Editor: Fasma Diele

Received: 5 March 2021
Accepted: 19 April 2021
Published: 20 April 2021

Publisher's Note: MDPI stays neutral with regard to jurisdictional claims in published maps and institutional affiliations.

Copyright: © 2021 by the authors. Licensee MDPI, Basel, Switzerland. This article is an open access article distributed under the terms and conditions of the Creative Commons Attribution (CC BY) license (https://creativecommons.org/licenses/by/4.0/).

1. Introduction

The investigation of stability properties plays a prominent role in the qualitative theory of fractional-order systems, similarly as in the case of the classical theory of integer-order dynamical systems [1,2]. The classical Hartman–Grobman linearisation theorem, which states that the local behavior of a dynamical system in a neighborhood of a hyperbolic equilibrium is qualitatively equivalent to the behavior of its linearisation near the equilibrium, is extended to the case of fractional-order systems as well [3–5]. Consequently, linear stability analysis is of fundamental importance in the investigation of fractional-order systems, and, in particular, stability properties of linear autonomous systems of fractional-order differential equations play a key role in this context.

For single-order systems of fractional differential equations (FDEs), namely systems in which the FDEs have the same fractional order, the most important theoretical result, which may now be considered classical, is Matignon's stability theorem [6], recently generalized in [7] for the case when the fractional order belongs to the interval $(0, 2)$.

Thus far, the investigation of stability properties of multi-order (incommensurate) fractional-order systems has unquestionably received less consideration. We refer to [8–11] for the stability analysis of incommensurate fractional-order systems with rational orders. Moreover, closely linked to this research topic, bounded input bounded output stability of systems with irrational transfer functions has been investigated in [12,13]. Very recently, the asymptotic properties of solutions of several classes of linear multi-order systems of fractional differential equations (such as systems with block triangular coefficient matrices) have been considered in [14].

The main difficulty in establishing necessary and sufficient conditions for the stability of multi-order linear systems of fractional differential equations (conceivably comparable to the classical Routh–Hurwitz conditions for integer-order systems) is due to the fact that a large number of parameters are involved: the system's coefficients, as well as multiple fractional orders. Undoubtedly, the complexity of the problem is positively correlated with the system's dimension.

The case of two-dimensional multi-order fractional-order systems has been fully investigated in [15–17]. On one hand, necessary and sufficient conditions for the asymptotic stability and instability of the fractional-order system have been obtained, in terms of the main diagonal elements and the determinant of the system's matrix, as well as the fractional orders of the Caputo derivatives. Moreover, necessary and sufficient fractional-order independent conditions have also been presented, in terms of the main diagonal elements and the determinant of the system's matrix, which guarantee the asymptotic stability or instability of the considered two-dimensional system, regardless of the choice of the fractional-orders considered in the system. These latter results prove to be especially useful in practical applications where the exact fractional orders of the Caputo derivatives are not precisely known.

It is important to note that multi-term fractional-order differential equations [18] and their qualitative properties are sharply linked to multi-order systems of fractional differential equations. We refer to [11] for a thorough presentation of the relationship between these two concepts. The investigation of stability properties of multi-term FDEs is so far limited to two-term and three-term fractional-order differential and difference equations, which have been recently studied in [19–22]. However, due to the increasing complexity of the problem, equations with four or more fractional terms have not yet been investigated.

This paper is organized as follows: Section 2 illustrates the statement of the problem and the main definitions. Due to the importance in the description of the solution of linear systems of FDEs, in Section 3, we provide a detailed description of the Mittag–Leffler function, of its derivatives and of the corresponding asymptotic behavior. Section 4 investigates the stability properties of single-order systems of FDEs, by presenting classical Matignon's theorem and some simulations illustrating the different stability behavior in dependence of the spectral properties of the matrix system. Stability analysis of multi-order systems is discussed in Section 5; since general results are far from being formulated in this case, we focus on some special cases and we separately investigate two-dimensional systems, higher dimensional systems with block-triangular structure, and higher dimensional systems with some special fractional orders. Some concluding remarks are hence provided in the concluding Section 6.

2. Preliminaries

Consider an n-dimensional fractional-order system with Caputo derivatives:

$$^C D^q \mathbf{y}(t) = f(t, \mathbf{y}) \tag{1}$$

with $\mathbf{q} = (q_1, q_2, ..., q_n) \in (0, 1]^n$, assuming that $f : [0, \infty) \times \mathbb{R}^n \to \mathbb{R}^n$ is a continuous function on its domain of definition, Lipschitz-continuous with respect to the second variable and $\mathbf{y} : [0, \infty) \to \mathbb{R}^n$ a vector-valued function. With $^C D^q \mathbf{y}(t)$, we denote the application of the Caputo derivative of order $0 < q_i \leq 1$ to each component $y_i(t)$ of $\mathbf{y}(t)$, namely

$$^C D^q \mathbf{y}(t) = \begin{pmatrix} ^C D^{q_1} y_1(t) \\ ^C D^{q_2} y_2(t) \\ \vdots \\ ^C D^{q_n} y_n(t) \end{pmatrix}, \quad ^C D^{q_i} y_i(t) := \frac{1}{\Gamma(1 - q_i)} \int_0^t (t - \tau)^{-q_i} y_i'(\tau) d\tau.$$

Existence and uniqueness of the solution of initial value problems associated with system (1) is ensured by Corollary 2.4 from [14].

Whenever $q_1 = q_2 = \ldots = q_n$, system (1) is said to be single-order; otherwise, the term multi-order will be used.

Let us further assume that $\mathbf{y} = 0$ is an equilibrium solution of system (1), i.e.,

$$f(t, 0) = 0 \quad \text{for any } t \geq 0.$$

Definition 1. *Let $\alpha > 0$ and denote by $\varphi(t, \mathbf{y}_0)$ the unique solution of (1) satisfying the initial condition $\mathbf{y}(0) = \mathbf{y}_0 \in \mathbb{R}^n$. Then:*

i. *the trivial solution of (1) is called stable if for any $\varepsilon > 0$ there exists $\delta = \delta(\varepsilon) > 0$ such that, for every $\mathbf{y}_0 \in \mathbb{R}^n$ satisfying $\|\mathbf{y}_0\| < \delta$, we have $\|\varphi(t, \mathbf{y}_0)\| \leq \varepsilon$ for any $t \geq 0$;*
ii. *the trivial solution of (1) is called asymptotically stable if it is stable and there exists $\rho > 0$ such that $\lim_{t \to \infty} \varphi(t, \mathbf{y}_0) = 0$ for $\|\mathbf{y}_0\| < \rho$;*
iii. *the trivial solution of (1) is called $\mathcal{O}(t^{-\alpha})$-asymptotically stable if it is stable and there exists $\rho > 0$ such that, for any $\|\mathbf{y}_0\| < \rho$, we have:*

$$\|\varphi(t, \mathbf{y}_0)\| = \mathcal{O}(t^{-\alpha}) \quad \text{as } t \to \infty.$$

Remark 1. *In the particular case of linear systems of fractional-order differential equations with constant coefficients, we say that the system is stable/asymptotically stable/unstable if and only if its trivial solution is stable/asymptotically stable/unstable.*

3. Mittag–Leffler Functions, Derivatives and Asymptotic Behavior

In the analysis of linear systems of FDEs, a crucial role is played by the Mittag–Leffler (ML) function [23]

$$E_{\alpha,\beta}(z) = \sum_{k=0}^{\infty} \frac{z^k}{\Gamma(\alpha k + \beta)}, \quad \alpha > 0, \quad z \in \mathbb{C}, \tag{2}$$

where $\Gamma(x) = \int_0^\infty t^{x-1} e^{-t} dt$ is the Euler–Gamma function. Since $\Gamma(k+1) = k!$, $k \in \mathbb{N}$, this function generalizes the exponential function when $\alpha = \beta = 1$, namely $E_{1,1}(z) = e^z$. When $\beta = 1$, the notation $E_\alpha(z) := E_{\alpha,1}(z)$ is preferred.

For the purposes of this paper (the reasons will be clearer later on), it is convenient to study and introduce a further generalization of the ML function.

3.1. The Prabhakar Function and Its Asymptotic Properties

For three real parameters α, β and γ, the three-parameter Mittag–Leffler (ML) function, also known as the Prabhakar function [24], is defined by its series representation

$$E_{\alpha,\beta}^{\gamma}(z) = \frac{1}{\Gamma(\gamma)} \sum_{k=0}^{\infty} \frac{\Gamma(\gamma + k) z^k}{k! \Gamma(\alpha k + \beta)}, \quad \alpha > 0, \quad z \in \mathbb{C}.$$

This function is not only a generalization, to three parameters, of the two-parameter ML function $E_{\alpha,\beta}(z)$ (indeed, when $\gamma = 1$, it is $E_{\alpha,\beta}^1(z) = E_{\alpha,\beta}(z)$), but it also provides a simple and elegant way to represent derivatives of two-parameter ML functions since

$$E_{\alpha,\beta}^{(m)}(z) := \frac{d^m}{dz^m} E_{\alpha,\beta}(z) = m! E_{\alpha,\alpha m+\beta}^{m+1}(z), \quad m = 0, 1, 2, \ldots, \tag{3}$$

as one can easily check after a term-by-term differentiation of (2).

In order to introduce a result about the Laplace transform (LT), it is necessary to introduce what is known as the *Prabhakar kernel*

$$e_{\alpha,\beta}^{\gamma}(t; \lambda) = t^{\beta-1} E_{\alpha,\beta}^{\gamma}(t^\alpha \lambda), \quad t > 0, \quad \lambda \in \mathbb{C},$$

for which the following analytical representation of the LT is available:

$$\mathcal{E}_{\alpha,\beta}^{\gamma}(s;\lambda) := \mathcal{L}\left[e_{\alpha,\beta}^{\gamma}(t;\lambda);s\right] = \frac{s^{\alpha\gamma-\beta}}{(s^{\alpha}-\lambda)^{\gamma}}, \quad \Re(s) > 0, \quad |s| > |\lambda|^{\frac{1}{\alpha}}.$$

Having in mind the stability analysis of linear FDEs, whose solutions will be expressed in terms of Mittag–Leffler functions and their derivatives, it is of interest to recall some results about the asymptotic behavior of the Prabhakar function in the complex plane.

In particular, for large arguments and $0 < \alpha \leq 1$, we first identify exponential and algebraic expansions, respectively given by

$$\mathcal{F}_{\alpha,\beta}^{\gamma}(z) = \frac{1}{\Gamma(\gamma)} e^{z^{1/\alpha}} z^{\frac{\gamma-\beta}{\alpha}} \frac{1}{\alpha^{\gamma}} \sum_{j=0}^{\infty} c_j z^{-\frac{j}{\alpha}}$$

$$\mathcal{A}_{\alpha,\beta}^{\gamma}(z) = \frac{z^{-\gamma}}{\Gamma(\gamma)} \sum_{j=0}^{\infty} \frac{(-1)^j \Gamma(j+\gamma)}{j!\Gamma(\beta-\alpha(j+\gamma))} z^{-j},$$

and, thanks to the results obtained by Paris [25,26], we know that

$$E_{\alpha,\beta}^{\gamma}(z) \sim \begin{cases} \mathcal{F}_{\alpha,\beta}^{\gamma}(z) + \mathcal{A}_{\alpha,\beta}^{\gamma}(ze^{\mp\pi i}) & |\arg z| < \frac{\alpha\pi}{2} \\ \mathcal{A}_{\alpha,\beta}^{\gamma}(ze^{\mp\pi i}) + \mathcal{F}_{\alpha,\beta}^{\gamma}(z) & \frac{\alpha\pi}{2} < |\arg z| < \alpha\pi \\ \mathcal{A}_{\alpha,\beta}^{\gamma}(ze^{\mp\pi i}) & \alpha\pi < |\arg z| \leq \pi \end{cases}, \quad |z| \to \infty$$

with the sign in $e^{\mp\pi i}$ which must taken negative for z in the upper complex half-plane and positive otherwise. Following the convention adopted in [27], in each sum, we have first indicated the dominant term, namely the exponential term $\mathcal{F}_{\alpha,\beta}^{\gamma}(z)$ when $|\arg z| < \frac{\alpha\pi}{2}$ and the algebraic term $\mathcal{A}_{\alpha,\beta}^{\gamma}(ze^{\mp\pi i})$ when $\frac{\alpha\pi}{2} < |\arg z| < \alpha\pi$. The lines $|\arg z| = \alpha\pi$ and $|\arg z| = \frac{\alpha\pi}{2}$ are, respectively, Stokes and anti-Stokes lines where asymptotic expansions change their behavior. The above result is graphically summarized in Figure 1.

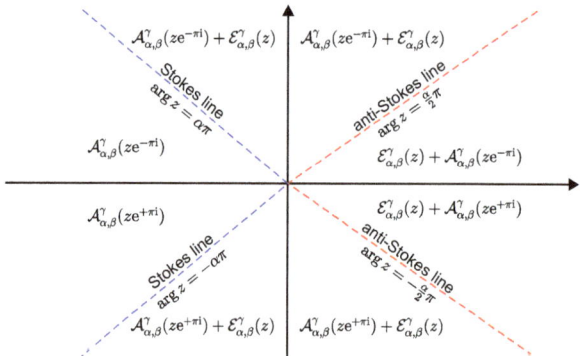

Figure 1. Asymptotic behavior of the Prabhakar function in the complex plane.

We first recall that coefficients c_j in asymptotic expansion $\mathcal{F}_{\alpha,\beta}^{\gamma}(z)$ are obtained [25] from the inverse factorial expansion, for $|s| \to \infty$ in $|\arg(s)| \leq \pi - \epsilon$ and any arbitrarily small $\epsilon > 0$, of

$$F_{\alpha,\beta}^{\gamma}(s) := \frac{\Gamma(\gamma+s)\Gamma(\alpha s + \psi)}{\Gamma(s+1)\Gamma(\alpha s + \beta)} = \alpha^{1-\gamma}\left(1 + \sum_{j=1}^{\infty} \frac{c_j}{(\alpha s + \psi)_j}\right) \quad (4)$$

with $(x)_j = \Gamma(x+j)/\Gamma(x)$ the Pochhammer symbol and $\psi = 1 - \gamma + \beta$. They can be evaluated by means of a sophisticated algorithm introduced in in [25] and also explained in [28]. The first few entries of c_k are available in [26].

Based on the asymptotic properties of the Prabhakar function, we obtain the asymptotic equivalence:

$$e_{\alpha,\beta}^{\gamma}(t;\lambda) \sim \begin{cases} \dfrac{\lambda^{\frac{\gamma-\beta}{\alpha}}}{\Gamma(\gamma)} t^{\gamma-1} e^{t\lambda^{1/\alpha}} & \text{if } |\arg(\lambda)| \leq \frac{\alpha\pi}{2} \\ \dfrac{e^{\pm\gamma\pi i}}{\lambda^{\gamma}\Gamma(\beta-\alpha\gamma)} t^{\beta-\alpha\gamma-1} & \text{if } |\arg(\lambda)| > \frac{\alpha\pi}{2} \end{cases} \quad (5)$$

as $t \to \infty$, where the sign in the term $e^{\pm\gamma\pi i}$ is positive if λ is in the upper complex half-plane, and negative otherwise.

3.2. Asymptotic Behavior of Derivatives of the ML Function

Thanks to the relationship (3) between derivatives of the ML function and the Prabhakar function, the investigation of the asymptotic behavior of any m-th order derivative of $E_{\alpha,\beta}(z)$ is hence possible by applying the corresponding results for the Prabhakar function and afterwards replacing β with $\alpha m + \beta$ and γ with $m+1$.

To this purpose, we first observe that, after these replacements, the function $F_{\alpha,\beta}^{\gamma}(s)$ in (4) becomes $F_{\alpha,\alpha m+\beta}^{m+1}(s) = (s+1)_m/(\alpha s + \psi)_m$, with $\psi = \alpha m + \beta$. Hence, coefficients c_j vanish for $j = m+1, m+2, \ldots$ and the exponential and algebraic expansions read

$$\mathcal{F}_{\alpha,\alpha m+\beta}^{m+1}(z) = \frac{1}{m!} e^{z^{1/\alpha}} z^{\frac{1-\alpha m-\beta}{\alpha}} \frac{1}{\alpha^{m+1}} \sum_{j=0}^{m} c_j z^{\frac{m-j}{\alpha}}$$

$$\mathcal{A}_{\alpha,\alpha m+\beta}^{m+1}(z) = \frac{1}{m!} \sum_{j=0}^{\infty} \frac{(-1)^j (j+1)_m}{\Gamma(\beta-\alpha(j+1))} z^{-j-m-1},$$

Therefore, by taking into account just the dominant expansions in each sector of the complex plane delimited by Stokes and anti-Stokes lines, and just leading terms in each expansion, we can describe the asymptotic behavior of derivatives of the ML function as

$$\frac{d^m}{dz^m} E_{\alpha,\beta}(z) \sim \begin{cases} \dfrac{1}{\alpha^{m+1}} e^{z^{1/\alpha}} z^{\frac{m+1-\alpha m-\beta}{\alpha}} & |\arg z| < \frac{\alpha\pi}{2} \\ (-1)^{m+1} \dfrac{m!}{\Gamma(\beta-1)} z^{-m-1} & \alpha\pi < |\arg z| \leq \pi \end{cases}, \quad |z| \to \infty$$

3.3. Behavior of Derivatives of the ML Function When $|\arg z| = \frac{\alpha\pi}{2}$

It remains to investigate the behavior along the anti-Stokes line $|\arg z| = \frac{\alpha\pi}{2}$ where both the exponential and the algebraic terms are present. We therefore consider

$$z = \rho e^{\pm\frac{\alpha\pi}{2}i}, \quad \rho > 0$$

and, for large $\rho = |z|$, it is

$$\frac{d^m}{dz^m} E_{\alpha,\beta}(z) \sim e^{\pm i\rho^{1/\alpha}} \frac{1}{\alpha^{m+1}} \sum_{j=0}^{m} c_j \rho^{\frac{1-\alpha m-\beta+m-j}{\alpha}} e^{\pm(1-\alpha m-\beta+m-j)\frac{\pi}{2}i} +$$

$$+ \sum_{j=0}^{\infty} \frac{(-1)^j (j+1)_m}{\Gamma(\beta-\alpha(j+1))} \frac{1}{\rho^{m+1+j}} e^{-i(m+1+j)(\frac{\alpha}{2}-1)\pi}.$$

Clearly, the second term asymptotically goes to zero when $\rho \to \infty$. The first term, instead, in modulus asymptotically tends to zero only for suitable values of α and β such that $1 - \alpha m - \beta + m - j \leq 0$ for any $j \in \{0, 1, \ldots, m\}$, namely, when $1 - \alpha m - \beta + m \leq 0$ or, equivalently, when

$$m \leq \frac{\beta - 1}{1 - \alpha}.$$

When we consider the one-parameter ML function $E_\alpha(z)$, namely $\beta = 1$, which is the instance of the ML involved in the stability analysis of linear FDEs, for $\arg z = \pm \frac{\alpha \pi}{2}$ just $|E_\alpha(z)|$ asymptotically converges to $1/\alpha$ for $|z| \to \infty$ but any m-th order derivative of $E_\alpha(z)$, with $m \geq 0$, is unbounded when $|z| \to \infty$.

This situation is illustrated in Figure 2, where we report the first derivatives of $E_\alpha(z)$ for $\alpha = 0.6$ and $\alpha = 0.8$ evaluated for z along the anti-Stokes line $\arg z = \frac{\alpha \pi}{2}$ (results are similar when $\arg z = -\frac{\alpha \pi}{2}$).

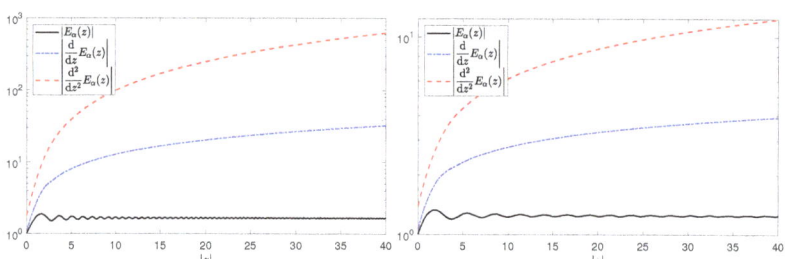

Figure 2. Modulus of $E_\alpha(z)$ and its first and second derivatives with $\arg z = \frac{\alpha \pi}{2}$ and $\alpha = 0.6$ (**left** plot) and $\alpha = 0.8$ (**right** plot).

Remark 2. *The behavior on the anti-Stokes line $|\arg z| = \frac{\alpha \pi}{2}$ of m-th order derivatives of the ML function $E_\alpha(z)$, which are unbounded as $|z| \to \infty$ for $m \geq 1$, is quite different from that of the exponential function e^z (namely, the special instance of $E_\alpha(z)$ for $\alpha = 1$). Indeed, derivatives of the the exponential are never unbounded on the corresponding anti-Stokes lines $|\arg z| = \frac{\pi}{2}$ since there it is $|d^m/dz^m e^z| = 1$ for any $m = 0, 1, \ldots$*

4. Stability of Linear Systems of Single-Order FDEs

We first consider the following linear system of Caputo-type fractional-order differential equations of the same fractional order:

$$^C D^q y(t) = A y(t), \tag{6}$$

where $q \in (0, 1]$ and $A \in \mathbb{R}^{n \times n}$, coupled with the initial condition $y(0) = y_0 \in \mathbb{R}^n$.

In is important to emphasize that system (6) is equivalent to the following system of weakly singular Volterra integral equations of convolution type (see, for example, [29,30]):

$$y(t) = y_0 + A \int_0^t \frac{(t - \tau)^{q-1}}{\Gamma(q)} y(\tau) d\tau. \tag{7}$$

For the most important advances regarding the general theory of linear Volterra integral equations, including the case when the convolution kernel is completely monotonic, we refer to [31–34].

The characteristic equation associated with system (6) is

$$\det(s^q I - A) = 0, \tag{8}$$

where, according to [35], the principal value (first branch) of the complex power function is considered. Therefore, it is easy to see that s is a root of the characteristic Equation (8) if and only if there exists an eigenvalue λ of the matrix A such that

$$s^q = \lambda. \tag{9}$$

Hence, this leads to the following characterization of the stability properties of system (6), in terms of the roots of its characteristic equation:

Proposition 1. *The linear system (6) is asymptotically stable if and only if*

$$\sigma(A) \subset S_q$$

where $\sigma(A)$ denotes the spectrum of the matrix A and

$$S_q = \{\lambda \in \mathbb{C} : s^q \neq \lambda, \forall \, \Re(s) \geq 0\}.$$

With the aim of investigating the stability properties of system (6) by characterizing the stability region S_q, and presenting a concise proof of Matignon's theorem [6], it is convenient to use the Jordan normal form of the matrix A. Indeed, let us consider a nonsingular matrix $P \in \mathbb{C}^{n \times n}$ such that

$$A = PJP^{-1}, \quad J = \begin{pmatrix} J_1 & 0 & \cdots & 0 \\ 0 & J_2 & \cdots & 0 \\ \vdots & \vdots & \ddots & \vdots \\ 0 & 0 & \cdots & J_p \end{pmatrix}$$

where J_k, $k = 1, \ldots, p$ are Jordan blocks

$$J_k = \begin{pmatrix} \lambda_k & 1 & 0 & \cdots & 0 & 0 \\ 0 & \lambda_k & 1 & \cdots & 0 & 0 \\ \vdots & \ddots & \ddots & \ddots & \ddots & \vdots \\ 0 & 0 & 0 & \cdots & \lambda_k & 1 \\ 0 & 0 & 0 & \cdots & 0 & \lambda_k \end{pmatrix}$$

and λ_k are eigenvalues of the matrix A. The size of the largest Jordan block J_k of A associated with the eigenvalue λ_k is called the *index* of λ_k [36]. On the other hand, the total number of Jordan blocks associated with a given eigenvalue λ_k in the Jordan normal form of the matrix A is the *geometric multiplicity* of the eigenvalue λ_k. Moreover, the sum of the sizes of all Jordan blocks corresponding to λ_k is the *algebraic multiplicity* of λ_k. Therefore, the index of an eigenvalue λ_k is equal to 1 if and only if its algebraic and geometric multiplicities are equal.

With these observations, we next give a slightly modified version of the classical result of Matignon, to fix a small imprecision in the second statement, related to the use of the *geometric multiplicity* instead of the *index* of an eigenvalue:

Theorem 1 (Matignon, 1996 [6]). *The linear system (6) is*

i. $\mathcal{O}(t^{-q})$-*asymptotically stable if and only if*

$$\sigma(A) \subset S_q = \left\{ \lambda \in \mathbb{C} : |\arg(\lambda)| > \frac{q\pi}{2} \right\}.$$

ii. *stable if and only if $\sigma(A) \subset \overline{S_q}$ and the eigenvalues of A which satisfy $|\arg(\lambda)| = \frac{q\pi}{2}$ have index 1.*

Proof. With the notations introduced previously, denoting $z(t) = Py(t)$, it is easy to verify that system (6) is equivalent to

$$^C D^q z(t) = Jz(t). \tag{10}$$

Applying the LT to the linear system (10) leads to the following formula for the LT of the vector function $z(t)$:

$$Z(s) = s^{q-1}(s^q I - J)^{-1} z(0) \tag{11}$$

Since the Jordan normal form J is a block diagonal matrix, the matrix $(s^q I - J)^{-1}$ is also block diagonal, and its blocks are upper triangular matrices of the form:

$$(s^q I - J_k)^{-1} = \begin{pmatrix} (s^q - \lambda_k)^{-1} & (s^q - \lambda_k)^{-2} & \cdots & (s^q - \lambda_k)^{-d_k} \\ 0 & (s^q - \lambda_k)^{-1} & \cdots & (s^q - \lambda_k)^{-(d_k-1)} \\ \vdots & \vdots & \ddots & \vdots \\ 0 & 0 & \cdots & (s^q - \lambda_k)^{-1} \end{pmatrix}$$

where d_k represents the dimension of the k-th Jordan block J_k.

Correspondingly, the Laplace transform $Z(s)$ is made up of "blocks" (of size d_k) of the form

$$Z_k(s) = s^{q-1}(s^q I - J_k)^{-1} z_k(0) = \begin{pmatrix} \sum_{j=1}^{d_k} \frac{s^{q-1}}{(s^q - \lambda_k)^j} z_{k,j}(0) \\ \sum_{j=2}^{d_k} \frac{s^{q-1}}{(s^q - \lambda_k)^{j-1}} z_{k,j}(0) \\ \vdots \\ \frac{s^{q-1}}{s^q - \lambda_k} z_{k,d_k}(0) \end{pmatrix}, \quad k = \overline{1, p}.$$

Applying the inverse LT, and taking into account that

$$\mathcal{L}^{-1}\left[\frac{s^{q-1}}{(s^q - \lambda_k)^m}; t\right] = t^{(m-1)q} E^m_{q,(m-1)q+1}(t^q \lambda_k) = e^m_{q,(m-1)q+1}(t; \lambda_k), \quad m \in \mathbb{N}^*$$

we obtain:

$$z_k(t) = \begin{pmatrix} \sum_{j=1}^{d_k} e^j_{q,(j-1)q+1}(t; \lambda_k) z_{k,j}(0) \\ \sum_{j=2}^{d_k} e^{j-1}_{q,(j-2)q+1}(t; \lambda_k) z_{k,j}(0) \\ \vdots \\ e^1_{q,1}(t; \lambda_k) z_{k,d_k}(0) \end{pmatrix}, \quad k = \overline{1, p}.$$

Based on (5), we obtain the following asymptotic equivalence:

$$e^m_{q,(m-1)q+1}(t; \lambda) = t^{(m-1)q} E^m_{q,(m-1)q+1}(t^q \lambda) \sim \begin{cases} \dfrac{\lambda^{(m-1)(\frac{1}{q}-1)}}{(m-1)! q^m} t^{m-1} e^{\lambda^{1/q} t} & \text{if } |\arg(\lambda)| \leq \frac{q\pi}{2} \\ \dfrac{(-1)^m}{\lambda^m \Gamma(1-q)} t^{-q} & \text{if } |\arg(\lambda)| > \frac{q\pi}{2} \end{cases}$$

as $t \to \infty$, where $m \in \mathbb{N}^*$.

Therefore, the following conclusions can be drawn:

- $e^m_{q,(m-1)q+1}(t; \lambda)$ converges to 0 as $t \to \infty$, if and only if $|\arg(\lambda)| > \frac{q\pi}{2}$; moreover, in this case, $e^m_{q,(m-1)q+1}(t; \lambda) = \mathcal{O}(t^{-q})$ as $t \to \infty$;
- if $|\arg(\lambda)| < \frac{q\pi}{2}$, the function $e^m_{q,(m-1)q+1}(t; \lambda)$ is unbounded;
- if $|\arg(\lambda)| = \frac{q\pi}{2}$, the function $e^m_{q,(m-1)q+1}(t; \lambda)$ is bounded if and only if $m = 1$.

With the above observations, the conclusions of Matignon's theorem readily follow. We emphasize that, for the case of statement ii., if there exists an eigenvalue of A which satisfies $|\arg(\lambda)| = \frac{q\pi}{2}$, the solutions of (6) are bounded if and only if the size of the largest Jordan block associated with this critical eigenvalue is equal to 1, i.e., the index of the eigenvalue is 1. □

Remark 3. The above proof slightly differs from the one in [6]. Matignon's proof, indeed, makes use of derivatives of the ML function instead of the Prabhakar kernel $e_{\alpha,\beta}^\gamma(t;\lambda)$ as in the proof of Theorem 1. A link between the two proofs can be, however, easily established in view of the relationship (3) between derivatives of the ML function and the Prabhakar function.

Remark 4. Matignon's theorem implies that, if $0 < q_1 < q_2 \leq 1$ and system (6) is asymptotically stable for $q = q_2$, then it will be asymptotically stable for $q = q_1$ as well. In particular, if the classical integer-order system $\dot{y} = Ay$ is asymptotically stable (i.e., all eigenvalues of A have negative real part), it follows that the fractional-order system (6) is asymptotically stable, for any fractional-order $q \in (0,1)$.

Example 1. To present numerical evidences of the above results, we consider here the linear systems of FDEs (6), with fractional order $q = 2/3$ and the coefficient matrix A chosen from one of the following four matrices:

$$A_1 = \begin{pmatrix} 1 & -\sqrt{3} & \frac{1}{4} & 0 \\ \sqrt{3} & 1 & 0 & \frac{1}{4} \\ 0 & 0 & 1 & -\sqrt{3} \\ 0 & 0 & \sqrt{3} & 1 \end{pmatrix}, \quad A_2 = \begin{pmatrix} 1 & -\sqrt{3} & 0 & 0 \\ \sqrt{3} & 1 & 0 & 0 \\ 0 & 0 & 1 & -\sqrt{3} \\ 0 & 0 & \sqrt{3} & 1 \end{pmatrix},$$

$$A_3 = \begin{pmatrix} 1-\epsilon & -\sqrt{3} & \frac{1}{4} & 0 \\ \sqrt{3} & 1-\epsilon & 0 & \frac{1}{4} \\ 0 & 0 & 1-\epsilon & -\sqrt{3} \\ 0 & 0 & \sqrt{3} & 1-\epsilon \end{pmatrix}, \quad A_4 = \begin{pmatrix} 1+\epsilon & -\sqrt{3} & 0 & 0 \\ \sqrt{3} & 1+\epsilon & 0 & 0 \\ 0 & 0 & 1+\epsilon & -\sqrt{3} \\ 0 & 0 & \sqrt{3} & 1+\epsilon \end{pmatrix}.$$

The solution $y(t) = E_q(t^q A)y_0$ evaluates by direct computation the matrix ML function thanks to the algorithm described in [37], after using the initial condition $y_0 = (1,-4,-2,4)^T$. The value $\epsilon = 0.1$ is used in A_3 and A_4.

The asymptotic behavior of the solution $y(t)$ depends on the spectral properties of the matrix. In particular, we observe that:

A_1 has two eigenvalues $\lambda_{1/2} = e^{\pm q\frac{\pi}{2}i}$ laying on the border of the stability sector S_q and both having index 2; according to Theorem 1, the system produces unbounded solutions as clearly shown in the left plot of Figure 3;

A_2 has the same two eigenvalues $\lambda_{1/2} = e^{\pm q\frac{\pi}{2}i}$ of A_1, laying on the border of the stability sector S_q, but their index is now 1; the expected bounded solutions are shown in the right plot of Figure 3;

A_3 has two eigenvalues $\lambda_{1/2}$ with index 2, as A_1, but now they lay inside the stability sector S_q; the asymptotically stable solutions are illustrated in the left plot of Figure 4;

A_4 has two eigenvalues $\lambda_{1/2}$ with index 1, as A_2, but lying outside the stability sector S_q; the resulting unbounded solutions are illustrated in the right plot of Figure 4.

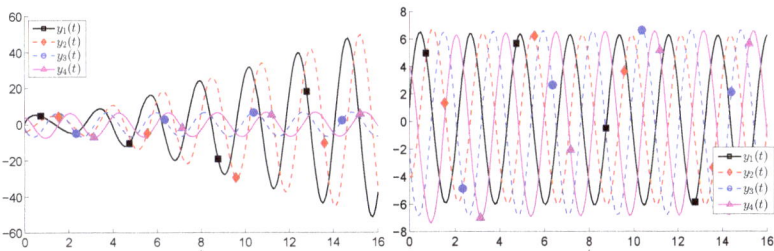

Figure 3. Solutions of the linear system ${}^C D_0^q = Ay(t)$, with $q = 2/3$, for $A = A_1$ (**left** plot) and $A = A_2$ (**right** plot).

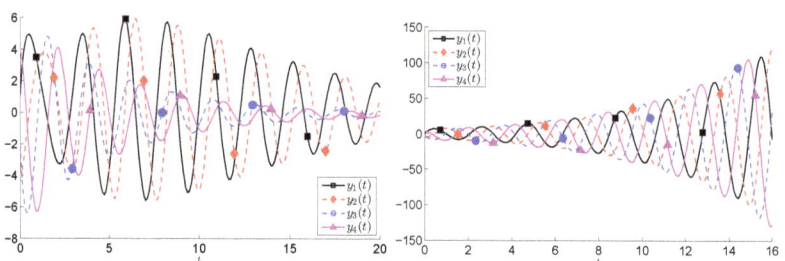

Figure 4. Solutions of the linear system $^C D_0^q = Ay(t)$, with $q = 2/3$, for $A = A_3$ (**left** plot) and $A = A_4$ (**right** plot).

5. Stability of Linear Multi-Order Systems of FDEs

Extending Matignon's theorem to the case of systems of FDEs with multiple fractional orders raises several technical difficulties, and, consequently, with the current state of the art, we are unable to present an exhaustive theory regarding this matter.

One of the technical difficulties that should be mentioned in this context is the fact that, for a general multi-order system of the form

$$^C D^q y(t) = Ay(t), \qquad (12)$$

where $A \in \mathbb{R}^{n \times n}$, $q = (q_1, q_2, \ldots, q_n) \in (0, 1]^n$ (such that not all q_i are equal), considering the Jordan normal form J of the matrix A and a nonsingular matrix $P \in \mathbb{C}^{n \times n}$ such that $A = PJP^{-1}$ (similarly as in the previous section), the transformation $z(t) = Py(t)$ does not lead to an equivalent system of the form

$$^C D^q z(t) = Jz(t).$$

Therefore, different theoretical approaches should be used to tackle linear multi-order systems of FDEs.

Using another approach, namely the Laplace transform method, we first obtain the following system:

$$s^{q_i} Y_i(s) - s^{q_i - 1} y_i(0) = \sum_{j=1}^{n} a_{ij} Y_j(s), \quad i = \overline{1,n}, \qquad (13)$$

where $Y_i(s)$ is the Laplace transform of the i-th component $y_i(t)$ of the solution $y(t)$.

System (13) is equivalent to the following system:

$$\Delta(s) \cdot \begin{pmatrix} Y_1(s) \\ Y_2(s) \\ \vdots \\ Y_n(s) \end{pmatrix} = \begin{pmatrix} b_1(s) \\ b_2(s) \\ \vdots \\ b_n(s) \end{pmatrix},$$

where $b_i(s) = s^{q_i - 1} y_i(0)$, for any $i = \overline{1,n}$ and

$$\Delta(s) = \text{diag}(s^{q_1}, s^{q_2}, \ldots, s^{q_n}) - A.$$

Using standard properties of the Laplace transform [8,14,35], the following result holds:

Theorem 2. *The multi-order system* (12) *is asymptotically stable if all the roots of the characteristic equation*

$$\det \Delta(s) = 0 \qquad (14)$$

have negative real parts.

It is important to point out that, for large-scale systems with many different fractional orders for the Caputo derivatives, the analysis of the roots of the characteristic Equation (14) is a very difficult and complex task.

Nevertheless, the case of two-dimensional linear multi-order systems has been fully analyzed in [17], and a summary of the main results will be presented in the next section.

5.1. Stability of Two-Dimensional Systems of FDEs with Different Fractional Orders

In the general case of a two-dimensional linear system of fractional-order differential equations:

$$\begin{cases} {}^{C}D^{q_1}y_1(t) = a_{11}y_1(t) + a_{12}y_2(t) \\ {}^{C}D^{q_2}y_2(t) = a_{21}y_1(t) + a_{22}y_2(t) \end{cases} \tag{15}$$

where $A = (a_{ij}) \in \mathbb{R}^{2\times 2}$ and $q_1, q_2 \in (0,1]$, applying the LT leads to the following characteristic equation:

$$\det(\mathrm{diag}(s^{q_1}, s^{q_2}) - A) = 0.$$

which can be written as

$$s^{q_1+q_2} - a_{11}s^{q_2} - a_{22}s^{q_1} + \det(A) = 0, \tag{16}$$

where s^{q_1} and s^{q_2} represent the principal values (first branches) of the corresponding complex power functions [35].

Employing asymptotic properties and the Final Value Theorem of the LT [12,35], the following result [16] holds:

Proposition 2.
1. System (15) is $\mathcal{O}(t^{-q})$-globally asymptotically stable (where $q = \min\{q_1, q_2\}$) if and only if all the roots (if any) of the characteristic Equation (16) are in the open left half-plane.
2. If $\det(A) \neq 0$ and the characteristic Equation (16) has a root in the open right half-plane, system (15) is unstable.

In general, computing the roots of the characteristic Equation (16) is not a straightforward task. Thus, departing from Proposition 2, we seek to obtain necessary and sufficient conditions involving the coefficients a_{11} and a_{22} of the main diagonal of the matrix A as well as the determinant $\det(A)$, which guarantee the stability or instability of system (15).

We first concentrate our attention on *fractional-order-dependent stability and instability conditions*, as described below. The proof of the following results is rather elaborate, involving the root locus method, and has been presented in detail in [17]. Note that only the case $\det(A) > 0$ is discussed here, as $\det(A) < 0$ implies that system (15) is unstable, for any fractional orders $(q_1, q_2) \in (0,1]^2$ (in fact, it is trivial to show that, if $\det(A) < 0$, the characteristic Equation (16) has at least one positive real root).

Lemma 1. Let $\delta > 0$, $q_1, q_2 \in (0,1]$ and consider the smooth parametric curve in the (a_{11}, a_{22})-plane defined by

$$\Gamma(\delta, q_1, q_2) : \begin{cases} a_{11} = \delta^{\frac{q_1}{q_1+q_2}} h(\omega, q_1, q_2) \\ a_{22} = \delta^{\frac{q_2}{q_1+q_2}} h(-\omega, q_1, q_2) \end{cases}, \quad \omega \in \mathbb{R},$$

where:

$$h(\omega, q_1, q_2) = \begin{cases} p_2(q_1, q_2)e^{q_1\omega} - p_1(q_1, q_2)e^{-q_2\omega}, & \text{if } q_1 \neq q_2 \\ \cos\frac{q\pi}{2} - \omega, & \text{if } q_1 = q_2 := q \end{cases}$$

with the functions $p_1(q_1, q_2)$ and $p_2(q_1, q_2)$ defined for $q_1 \neq q_2$ as

$$p_k(q_1, q_2) = \frac{\sin \frac{q_k \pi}{2}}{\sin \frac{(q_2 - q_1)\pi}{2}}, \quad \text{for } k = \overline{1,2}.$$

The following statements hold:

i. The curve $\Gamma(\delta, q_1, q_2)$ is the graph of a smooth, decreasing, concave bijective function $\phi_{\delta, q_1, q_2} : \mathbb{R} \to \mathbb{R}$ in the (a_{11}, a_{22})-plane.

ii. The curve $\Gamma(\delta, q_1, q_2)$ lies outside the third quadrant of the (a_{11}, a_{22})-plane.

Theorem 3 (Fractional-order-dependent stability and instability results).

Let $\det(A) = \delta > 0$ and $q_1, q_2 \in (0,1]$ arbitrarily fixed. Consider the curve $\Gamma(\delta, q_1, q_2)$ and the function $\phi_{\delta, q_1, q_2} : \mathbb{R} \to \mathbb{R}$ given by Lemma 1.

i. The characteristic Equation (16) has a pair of pure imaginary roots if and only if $(a_{11}, a_{22}) \in \Gamma(\delta, q_1, q_2)$.

ii. System (15) is $\mathcal{O}(t^{-q})$-asymptotically stable (with $q = \min\{q_1, q_2\}$) if and only if

$$a_{22} < \phi_{\delta, q_1, q_2}(a_{11}).$$

iii. If $a_{22} > \phi_{\delta, q_1, q_2}(a_{11})$, system (15) is unstable.

Theorem 3 provides a relatively simple algebraic criterion (in the form of inequalities comprising the elements of the main diagonal of the system's matrix A as well as its determinant and the fractional orders) that enables us to immediately decide the question of asymptotic stability or instability for a given two-dimensional multi-order system of fractional differential equations. In fact, Theorem 3 may be seen as a generalization of the Routh–Hurwitz stability criterion.

Remark 5. If $q_1 = q_2 := q$, the curve $\Gamma(\delta, q_1, q_2)$ reduces to the straight line:

$$a_{11} + a_{22} = 2\sqrt{\delta} \cos \frac{q\pi}{2}.$$

Therefore, Theorem 3 provides that, for equal fractional orders, system (15) is asymptotically stable if and only if

$$Tr(A) < 2\sqrt{\det(A)} \cos \frac{q\pi}{2}. \qquad (17)$$

The eigenvalues of the system's matrix A are

$$\lambda_{1,2} = \frac{Tr(A) \pm \sqrt{Tr(A)^2 - 4\det(A)}}{2}$$

and, hence, inequality (17) is equivalent to $|\arg \lambda_{1,2}| > \frac{q\pi}{2}$. Consequently, for two-dimensional systems, the conclusion of Matignon's theorem is recovered as a particular case of Theorem 3.

Remark 6. The asymptotic stability of the two-dimensional integer order system $\dot{y} = Ay$ does not directly imply the asymptotic stability of system (15) for any fractional orders $(q_1, q_2) \in (0,1]^2$. We can only state, based on Remark 4, that, if the integer order system $\dot{y} = Ay$ is asymptotically stable, then so is system (15) with equal fractional orders $q_1 = q_2$.

Example 2. Let us consider the system

$$\begin{cases} {}^C D^{q_1} y_1(t) = a_{11} y_1(t) + a_{12} y_2(t) \\ {}^C D^{q_2} y_2(t) = a_{21} y_1(t) + a_{22} y_2(t) \end{cases} \text{ with } A = (a_{ij}) = \begin{pmatrix} -2 & 0.5 \\ -5 & 1 \end{pmatrix} \qquad (18)$$

where $q_1, q_2 \in (0,1]$. As $Tr(A) = -1 < 0$ and $\det(A) = 0.5 > 0$, the Routh–Hurwith stability test guarantees that, for $q_1 = q_2 = 1$, system (18) is asymptotically stable (see left plot in Figure 5);

the eigenvalues of the matrix A are $\lambda_{1,2} = -\frac{1}{2}(1 \pm i)$. Therefore, for equal fractional orders $q_1 = q_2 \in (0,1)$, system (18) is also asymptotically stable (see left plot in Figure 5); this can also be verified by inequality (17).

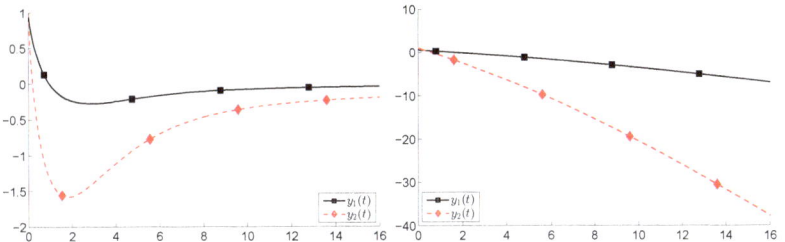

Figure 5. Asymptotically stable solutions of system (18) when $q = (0.8, 0.8)$ (**left** plot) and unstable solutions when $q = (0.2, 1)$ (**right** plot).

However, for $q_1 = 0.2$ and $q_2 = 1$, system (18) is unstable (see right plot in Figure 5). Indeed, applying Theorem 3, system (18) with $(q_1, q_2) = (0.2, 1)$ is unstable if

$$a_{22} > \phi_{\delta,q_1,q_2}(a_{11}),$$

where $a_{11} = -2$, $a_{22} = 1$, $\delta = \det(A) = 0.5$ and, based on the notations from Lemma 1:

$$\phi_{\delta,q_1,q_2}(a_{11}) = \delta^{\frac{q_2}{q_1+q_2}} h(-\omega^*, q_1, q_2)$$

where ω^* is the unique root of the equation

$$a_{11} = \delta^{\frac{q_1}{q_1+q_2}} h(\omega^*, q_1, q_2).$$

Numerically solving this algebraic equation, we compute $\omega^* = -2.19664$ and, therefore, we also obtain $\phi_{\delta,q_1,q_2}(a_{11}) = 0.895383$. As $a_{22} = 1$, it follows that the instability condition $a_{22} > \phi_{\delta,q_1,q_2}(a_{11})$ is satisfied (see left plot of Figure 6).

Furthermore, it is important to emphasize that Theorem 3 can also be applied when at least one of the fractional orders is irrational. For example, choosing $q_1 = \frac{1}{\pi}$ and $q_2 = 1$, in a similar way as before, we compute $\phi_{\delta,q_1,q_2}(a_{11}) = 1.10307$, and hence $a_{22} < \phi_{\delta,q_1,q_2}(a_{11})$, which means that system (18) with $q_1 = \frac{1}{\pi}$ and $q_2 = 1$ is asymptotically stable (see right plot of Figure 6).

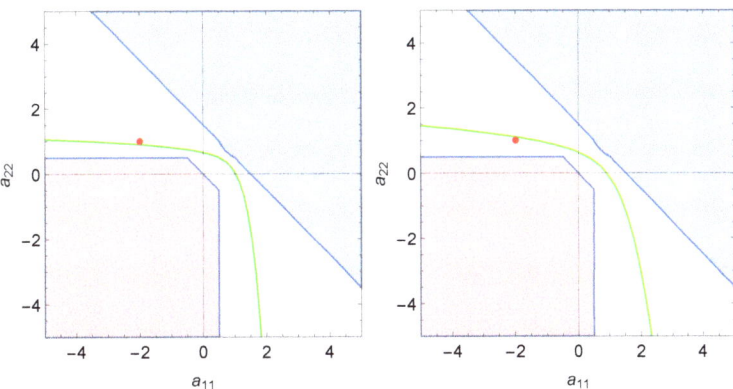

Figure 6. Position of the point $(a_{11}, a_{22}) = (-2, 1)$ (plotted in red) with respect to curve $\Gamma(\delta, q_1, q_2)$ (shown in green) in the particular case $(q_1, q_2) = (0.2, 1)$ (**left** plot) and $(q_1, q_2) = \left(\frac{1}{\pi}, 1\right)$ (**right** plot) from Example 2.

Figure 7 showes region of fractional orders (q_1, q_2) for which system (18) is globally asymptotically stable.

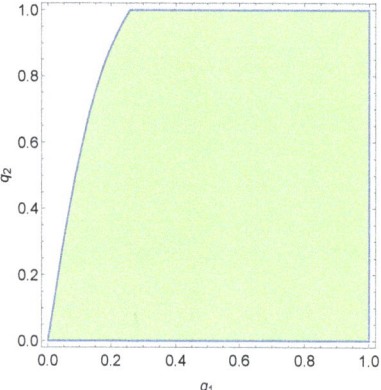

Figure 7. Region of fractional orders (q_1, q_2) for which system (18) is globally asymptotically stable.

The next step is to seek necessary and sufficient conditions which ensure the asymptotic stability or instability of system (15) for any choice of the fractional orders. A complete investigation of the family of curves $\Gamma(\delta, q_1, q_2)$ leads to the following *fractional-order independent stability and instability results* [17]:

Theorem 4 (Fractional-order independent instability results).
i. If $\det(A) < 0$, system (15) is unstable, regardless of the fractional orders q_1 and q_2.
ii. If $\det(A) > 0$, system (15) is unstable regardless of the fractional orders q_1 and q_2 if and only if one of the following conditions holds:

$$\begin{cases} a_{11} + a_{22} \geq \det(A) + 1 \text{ or} \\ a_{11} > 0,\ a_{22} > 0,\ a_{11}a_{22} \geq \det(A). \end{cases}$$

Theorem 5 (Fractional-order-independent stability results). *System (15) is asymptotically stable, regardless of the fractional orders $q_1, q_2 \in (0, 1]$ if and only if the following inequalities are satisfied:*

$$a_{11} + a_{22} < 0 < \det(A) \quad \text{and} \quad \max\{a_{11}, a_{22}\} < \min\{1, \det(A)\}.$$

The previous theorems provide easily verifiable necessary and sufficient conditions which ensure the asymptotic stability or instability of the two-dimensional system (15), for any choice of the fractional orders $q_1, q_2 \in (0, 1]$. These conditions are expressed as simple inequalities involving the main diagonal elements a_{11} and a_{22} as well as the determinant $\det(A)$ of the system's matrix. On one hand, if $\det(A) < 0$, Theorem 4 provides that system (15) is unstable, for any choice of the fractional orders $q_1, q_2 \in (0, 1]$. Hence, we will focus our attention on the case $\delta = \det(A) > 0$. Let us denote by R_s and by R_u the *fractional-order independent stability and instability regions* given by Theorems 4 and 5:

$$R_u = \{(a_{11}, a_{22}, \delta) \in \mathbb{R}^2 \times (0, \infty) : a_{11} + a_{22} \geq \delta + 1 \text{ or } a_{11} > 0,\ a_{22} > 0,\ a_{11}a_{22} \geq \delta\}$$
$$R_s = \{(a_{11}, a_{22}, \delta) \in \mathbb{R}^2 \times (0, \infty) : a_{11} + a_{22} < 0 \text{ and } \max\{a_{11}, a_{22}\} < \min\{1, \delta\}\}$$

The regions R_u and R_s are plotted in Figure 8. The intersections of these regions with the $\delta = \det(A) = 6$ plane are shown in Figure 9. Moreover, it can be verified [17] that the union of all the curves $\Gamma(\delta, q_1, q_2)$ (for $\delta > 0$ and $q_1, q_2 \in (0, 1]$) represents the complementary of $R_s \cup R_u$ (see Figure 9).

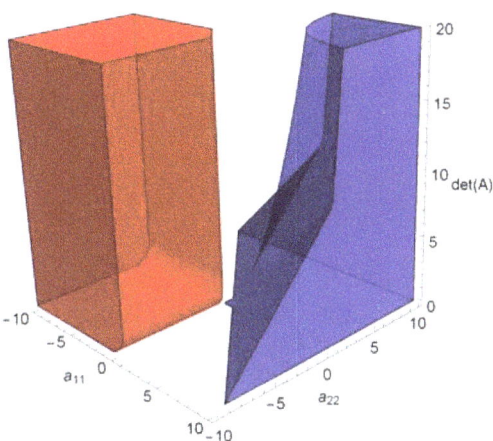

Figure 8. The *fractional-order-independent* stability (red) and instability (blue) regions R_s and R_u provided by Theorems 4 and 5 for system (15).

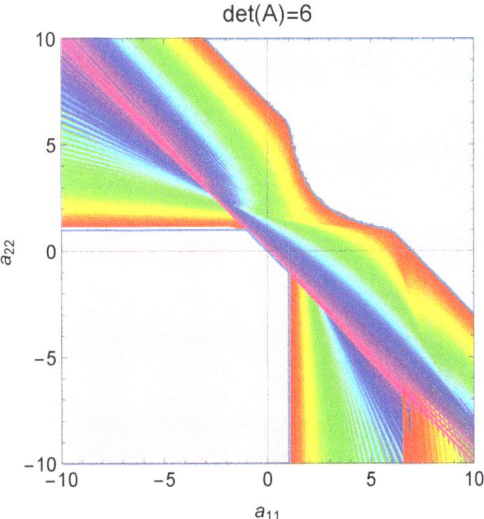

Figure 9. Curves $\Gamma(\delta, q_1, q_2)$ given by Lemma 1, for $\det(A) = \delta = 6$ and $q_i \in \left\{\frac{k}{40}, k = \overline{1,40}\right\}, i = \overline{1,2}$ (1600 curves), color-coded from red to violet according to increasing values of $q_1 q_2$. The red/blue shaded regions represent the intersections of the *fractional-order independent* stability and instability regions (see Figure 8) with the $\det(A) = 6$ plane.

Remark 7. *The classical Routh–Hurwitz stability test for two-dimensional systems of the form $\dot{y} = Ay$ provide that the system is asymptotically stable if and only if $Tr(A) < 0$ and $\det(A) > 0$. However, based on Theorem 5, the additional inequality $\max\{a_{11}, a_{22}\} < \min\{1, \det(A)\}$ has to be verified in order to ensure the asymptotic stability of the fractional-order system (15), regardless of the choice of fractional orders q_1 and q_2.*

Example 3. *It is easy to see that, if system (18) is considered, as $a_{11} = -2$, $a_{22} = 1$ and $\det(A) = 0.5$, even though the Routh–Hurwitz conditions $Tr(A) < 0$ and $\det(A) > 0$ are fulfilled, the additional additional inequality $\max\{a_{11}, a_{22}\} < \min\{1, \det(A)\}$ does not hold, and hence* system (18) is not asymptotically stable for any choice of the fractional orders $q_1, q_2 \in (0, 1]$. Indeed, as we have seen in Example 2, for $q_1 = 0.2$ and $q_2 = 1$, system (18) is unstable.

In conclusion, based on the previously described results, the following steps should be undertaken for the stability analysis of a two-dimensional system of FDEs:
1. if $\det(A) < 0$, then the system is unstable, for any choice of the fractional orders $q_1, q_2 \in (0, 1]$, based on Theorem 4;
2. if $(a_{11}, a_{22}, \det(A)) \in R_u$, then the system is unstable, for any choice of the fractional orders $q_1, q_2 \in (0, 1]$, based on Theorem 4;
3. if $(a_{11}, a_{22}, \det(A)) \in R_s$, then the system is asymptotically stable, for any choice of the fractional orders $q_1, q_2 \in (0, 1]$, based on Theorem 5;
4. if $(a_{11}, a_{22}, \det(A)) \notin R_s \cup R_u$, then the stability properties of the system depend on choice of the fractional orders $q_1, q_2 \in (0, 1]$ and Theorem 3 should be applied.

The results described in this section, particularly Theorems 3–5, give a comprehensive method to assess stability properties of two-dimensional fractional-order systems. However, the generalization of these results to higher dimensional fractional-order systems still remains an open question.

5.2. Stability of Higher Dimensional Systems of FDEs with Specific Structures

Consider that the matrix A of the linear system (12) has a block-triangular structure:

$$A = \begin{pmatrix} A_{11} & A_{12} & \cdots & A_{1p} \\ & A_{22} & \cdots & A_{2p} \\ & & \ddots & \vdots \\ & & & A_{pp} \end{pmatrix}$$

where $A_{ii} \in \mathbb{R}^{d_i \times d_i}$, for $i = \overline{1, m}$ and $A_{ii} \in \mathbb{R}^{2 \times 2}$ for $i = \overline{m+1, p}$, such that

$$\sum_{i=1}^{m} d_i + 2(p - m) = n.$$

We also assume that

$$q = (\underbrace{q_1, q_1, \ldots, q_1}_{d_1 \text{ times}}, \ldots, \underbrace{q_m, q_m, \ldots, q_m}_{d_m \text{ times}}, q_{m+1}^1, q_{m+1}^2, \ldots, q_p^1, q_p^2) \in (0, 1]^n.$$

In this case, the characteristic equation associated with system (12) is

$$\prod_{i=1}^{m} \det(s^{q_i} I - A_{ii}) \cdot \prod_{i=m+1}^{p} \det(\operatorname{diag}(s^{q_i^1}, s^{q_i^2}) - A_{ii}) = 0 \qquad (19)$$

Therefore, combining Matignon's theorem (Theorem 1) and Theorem 3, the following statements are obtained:
- system (12) is asymptotically stable if and only if
  - $\sigma(A_{ii}) \subset S_{q_i} = \{\lambda \in \mathbb{C} : |\arg(\lambda)| > \frac{q_i \pi}{2}\}$ for any $i = \overline{1, m}$ and
  - $A_{ii}^{22} < \phi_{\delta_i, a_i^1, a_i^2}(A_{ii}^{11})$, for any $i = \overline{m+1, p}$, where A_{ii}^{11} and A_{ii}^{22} are the main diagonal elements of matrix A_{ii}, $\delta_i = \det(A_{ii})$ and ϕ is defined in Lemma 1.
- system (12) is unstable if at least one of the following holds:
  - there exists $i \in \{1, 2, \ldots m\}$ such that the matrix A_{ii} has at least one eigenvalue λ such that $|\arg(\lambda)| < \frac{q_i \pi}{2}$ or

- there exists $i \in \{m+1,\ldots,p\}$ such that $A_{ii}^{22} > \phi_{\delta_i, q_i^1, q_i^2}(A_{ii}^{11})$, where A_{ii}^{11} and A_{ii}^{22} are the main diagonal elements of matrix A_{ii}, $\delta_i = \det(A_{ii})$ and ϕ is defined in Lemma 1.

5.3. Stability of Higher Dimensional Systems of FDEs with Special Fractional Orders

Let us consider the following n-dimensional linear multi-order system of fractional differential equations:

$$^C D^{q_i} y_i(t) = \sum_{j=1}^{n} a_{ij} y_j(t), \quad i = \overline{1, n}, \tag{20}$$

where $q_i \in (0,1]$, $a_{ij} \in \mathbb{R}$ and $n \geq 3$.

If the coefficient matrix of the system is not of a triangular or block triangular form as considered in the previous section, one can not provide a comprehensive stability theory. Still, an approach that works under certain restrictions on the fractional orders of the Caputo derivatives has been developed in [14]. We will next recall the general results obtained by the mentioned authors.

Suppose $q_j \in (0,1]$, for any $j = \overline{1,n}$ and that there exists $q^* \in (0,1]$ and $\rho_j \in \mathbb{Q}$ such that $q_j = \rho_j q^*$. It follows that there exists $r_j, s_j \in \mathbb{N}$ for $j = \overline{1,n}$ such that $\gcd(r_j, s_j) = 1$ and $\rho_j = \dfrac{r_j}{s_j}$. Let s be the least common multiple of the denominators s_j. Then, for any j, there exists $\alpha_j \in \mathbb{N}$ such that

$$q_j = \frac{q^* \alpha_j}{s} \quad \left(\alpha_j = \frac{s r_j}{s_j}\right).$$

We can rewrite the j-th equation of system (20) as an equivalent system of α_j differential equations having the order $\dfrac{q^*}{s}$. It follows that system (20) can be expressed as a system of $n^* = \sum_{j=1}^{n} \alpha_j$ equations of order $\dfrac{q^*}{s}$:

$$^C D^{q^*/s} y^*(t) = A^* y^*(t), \tag{21}$$

where A^* has the following block structure

$$A^* = \begin{pmatrix} A_{11} & A_{12} & \cdots & A_{1n} \\ A_{21} & A_{22} & \cdots & A_{2n} \\ \vdots & \vdots & \ddots & \vdots \\ A_{n1} & A_{n2} & \cdots & A_{nn} \end{pmatrix}$$

with $A_{jk} \in \mathbb{R}^{\alpha_j \times \alpha_k}$,

$$A_{jj} = \begin{pmatrix} 0 & 1 & 0 & \cdots & 0 & 0 \\ 0 & 0 & 1 & \cdots & 0 & 0 \\ \vdots & \vdots & \vdots & \ddots & 0 & 0 \\ 0 & 0 & 0 & \cdots & 0 & 1 \\ a_{jj} & 0 & 0 & \cdots & 0 & 0 \end{pmatrix}, j = \overline{1,n}$$

and

$$A_{jk} = \begin{pmatrix} 0 & 0 & \cdots & 0 \\ \vdots & \vdots & \ddots & 0 \\ 0 & 0 & \cdots & 0 \\ a_{jk} & 0 & \cdots & 0 \end{pmatrix}, j,k = \overline{1,n}, j \neq k.$$

Even though the dimension n^* of the system may be significantly higher than the dimension n of the original system (20), resulting in higher computational costs, all the equations of the new system (21) now have the same fractional order, giving an advantage in studying the stability properties of the solutions of the system.

We expose the main result of this section, based on [14], which gives us stability criteria involving the components of the matrix A^*.

Theorem 6. *Suppose that $q_j \in (0,1]$ for any j and there exists $q^* \in (0,1]$ and $\rho_j \in \mathbb{Q}$ such that $q_j = \rho_j q^*$, for all j. Then, all the solutions of system (20) converge to zero at infinity if the eigenvalues λ_j^* of the associated system's coefficient matrix A^* satisfy*

$$|\arg \lambda_j^*| > \frac{\pi q^*}{2s}, \; \forall j = \overline{1,n},$$

with s being the least common multiple of the denominators of ρ_j.

Example 4. *Again, we reconsider system (18) with $q_1 = 0.2$ and $q_2 = 1$. In this case, the matrix A^* given by the above procedure is*

$$A^* = \begin{pmatrix} a_{11} & a_{12} & 0 & 0 & 0 & 0 \\ 0 & 0 & 1 & 0 & 0 & 0 \\ 0 & 0 & 0 & 1 & 0 & 0 \\ 0 & 0 & 0 & 0 & 1 & 0 \\ 0 & 0 & 0 & 0 & 0 & 1 \\ a_{21} & a_{22} & 0 & 0 & 0 & 0 \end{pmatrix} = \begin{pmatrix} -2 & 0.5 & 0 & 0 & 0 & 0 \\ 0 & 0 & 1 & 0 & 0 & 0 \\ 0 & 0 & 0 & 1 & 0 & 0 \\ 0 & 0 & 0 & 0 & 1 & 0 \\ 0 & 0 & 0 & 0 & 0 & 1 \\ -5 & 1 & 0 & 0 & 0 & 0 \end{pmatrix}$$

and the system (18) is equivalent to a system of six fractional-order differential equations with the same order $q = q_1 = 0.2$. The matrix A^* has a pair of complex conjugated eigenvalues $(\lambda, \overline{\lambda})$, $\lambda = 0.543842 + i\, 0.133131$ such that $|\arg(\lambda)| = 0.240076 < 0.1\pi = \frac{q\pi}{2}$. Hence, based on Matignon's theorem, system (18) is unstable for $q_1 = 0.2$ and $q_2 = 1$. Therefore, this is in good agreement with the results obtained in Example 2, based on Theorem 3.

However, it is important to note that cases $q_1 = \frac{1}{\pi}$ and $q_2 = 1$ cannot be investigated using the technique provided by Theorem 6.

6. Conclusions

An extensive analysis of stability properties of linear systems of FDEs has been provided. This analysis is of importance to describe the asymptotic behavior of physical systems when modeled by means of FDEs. Both single-order and multi-order systems have been studied, reviewing the most important theoretical results that have been obtained so far in the literature. The role of the Mittag–Leffler function, and of its derivatives, has been highlighted and a presentation of their asymptotic behavior has been proposed. We have seen that, unlike systems of integer order, coefficients of the systems are not sufficient to describe stability properties of solutions, due to the tight dependence on the order of the fractional derivatives. This dependence becomes more and more difficult to investigate in systems incorporating derivatives of different order, as we have observed from the analysis of two-dimensional systems. Stability analysis of multi-order higher dimensional systems is still an open problem which deserves to be investigated with more attention; with this work, a first contribution has been provided by examining systems with some specific structures, and we hope these results will stimulate the analysis of more general systems.

Author Contributions: Conceptualization, O.B., R.G., and E.K.; methodology, O.B. and E.K.; software, O.B., R.G., and E.K.; validation, O.B., R.G., and E.K.; formal analysis, O.B. and E.K.; investigation, O.B., E.K., and R.G.; writing—original draft preparation, O.B., R.G., and E.K.; writing—review and editing, O.B., R.G., and E.K.; visualization, O.B., E.K., and R.G.; supervision, E.K.; funding acquisition, R.G. and E.K. All authors have read and agreed to the published version of the manuscript.

Funding: This research was funded by COST Action CA 15225—"Fractional-order systems-analysis, synthesis and their importance for future design". The work of R.G. has also been partially supported by an INdAM-GNCS 2020 project.

Institutional Review Board Statement: Not applicable.

Informed Consent Statement: Not applicable.

Data Availability Statement: Not applicable.

Conflicts of Interest: The authors declare no conflict of interest.

Abbreviations

The following abbreviations are used in this manuscript:

FDE   Fractional differential equation
ML   Mittag–Leffler
LT   Laplace transform

References

1. Li, C.; Zhang, F. A survey on the stability of fractional differential equations. *Eur. Phys. J. Spec. Top.* **2011**, *193*, 27–47. [CrossRef]
2. Rivero, M.; Rogosin, S.V.; Tenreiro Machado, J.A.; Trujillo, J.J. Stability of fractional order systems. *Math. Probl. Eng.* **2013**, *2013*, 356215.
3. Li, C.; Ma, Y. Fractional dynamical system and its linearization theorem. *Nonlinear Dyn.* **2013**, *71*, 621–633. [CrossRef]
4. Wang, Z.; Yang, D.; Zhang, H. Stability analysis on a class of nonlinear fractional-order systems. *Nonlinear Dyn.* **2016**, *86*, 1023–1033. [CrossRef]
5. Tuan, H.T.; Trinh, H. Global attractivity and asymptotic stability of mixed-order fractional systems. *IET Control Theory Appl.* **2020**, *14*, 1240–1245. [CrossRef]
6. Matignon, D. Stability results for fractional differential equations with applications to control processing. *Comput. Eng. Syst. Appl.* **1996**, *2*, 963–968.
7. Sabatier, J.; Farges, C. On stability of commensurate fractional order systems. *Int. J. Bifurc. Chaos* **2012**, *22*, 1250084. [CrossRef]
8. Deng, W.; Li, C.; Lu, J. Stability analysis of linear fractional differential system with multiple time delays. *Nonlinear Dyn.* **2007**, *48*, 409–416. [CrossRef]
9. Deng, W.; Li, C.; Guo, Q. Analysis of fractional differential equations with multi-orders. *Fractals* **2007**, *15*, 173–182. [CrossRef]
10. Petras, I. Stability of fractional-order systems with rational orders. *Fract. Calc. Appl. Anal.* **2009**, *12*, 269–298.
11. Diethelm, K. Multi-term fractional differential equations, multi-order fractional differential systems and their numerical solution. *J. Eur. Syst. Autom.* **2008**, *42*, 665–676. [CrossRef]
12. Bonnet, C.; Partington, J.R. Coprime factorizations and stability of fractional differential systems. *Syst. Control Lett.* **2000**, *41*, 167–174. [CrossRef]
13. Trächtler, A. On BIBO stability of systems with irrational transfer function. *arXiv* **2016**, arXiv:1603.01059.
14. Diethelm, K.; Siegmund, S.; Tuan, H. Asymptotic behavior of solutions of linear multi-order fractional differential systems. *Fract. Calc. Appl. Anal.* **2017**, *20*, 1165–1195. [CrossRef]
15. Brandibur, O.; Kaslik, E. Stability properties of a two-dimensional system involving one Caputo derivative and applications to the investigation of a fractional-order Morris-Lecar neuronal model. *Nonlinear Dyn.* **2017**, *90*, 2371–2386. [CrossRef]
16. Brandibur, O.; Kaslik, E. Stability of two-component incommensurate fractional-order systems and applications to the investigation of a FitzHugh-Nagumo neuronal model. *Math. Methods Appl. Sci.* **2018**, *41*, 7182–7194. [CrossRef]
17. Brandibur, O.; Kaslik, E. Exact stability and instability regions for two-dimensional linear autonomous multi-order systems of fractional-order differential equations. *Fract. Calc. Appl. Anal.* **2021**, *24*, 225–253. [CrossRef]
18. Atanackovic, T.; Dolicanin, D.; Pilipovic, S.; Stankovic, B. Cauchy problems for some classes of linear fractional differential equations. *Fract. Calc. Appl. Anal.* **2014**, *17*, 1039–1059. [CrossRef]
19. Jiao, Z.; Chen, Y.Q. Stability of fractional-order linear time-invariant systems with multiple noncommensurate orders. *Comput. Math. Appl.* **2012**, *64*, 3053–3058. [CrossRef]
20. Čermák, J.; Kisela, T. Asymptotic stability of dynamic equations with two fractional terms: continuous versus discrete case. *Fract. Calc. Appl. Anal.* **2015**, *18*, 437. [CrossRef]
21. Čermák, J.; Kisela, T. Stability properties of two-term fractional differential equations. *Nonlinear Dyn.* **2015**, *80*, 1673–1684. [CrossRef]
22. Brandibur, O.; Kaslik, E. Stability analysis of multi-term fractional-differential equations with three fractional derivatives. *J. Math. Anal. Appl.* **2021**, *495*, 124751. [CrossRef]
23. Haubold, H.; Mathai, A.; Saxena, R. Mittag–Leffler Functions and Their Applications. *J. Appl. Math.* **2011**, *2011*, 298628.
24. Prabhakar, T.R. A singular integral equation with a generalized Mittag–Leffler function in the kernel. *Yokohama Math. J.* **1971**, *19*, 7–15.
25. Paris, R.B. Exponentially small expansions in the asymptotics of the Wright function. *J. Comput. Appl. Math.* **2010**, *234*, 488–504. [CrossRef]

26. Paris, R.B. Asymptotics of the special functions of fractional calculus. In *Handbook of Fractional Calculus with Applications*; De Gruyter: Berlin, Germany, 2019; Volume 1, pp. 297–325.
27. Giusti, A.; Colombaro, I.; Garra, R.; Garrappa, R.; Polito, F.; Popolizio, M.; Mainardi, F. A practical guide to Prabhakar fractional calculus. *Fract. Calc. Appl. Anal.* **2020**, *23*, 9–54. [CrossRef]
28. Garra, R.; Garrappa, R. The Prabhakar or three parameter Mittag-Leffler function: theory and application. *Commun. Nonlinear Sci. Numer. Simul.* **2018**, *56*, 314–329. [CrossRef]
29. Giusti, A. General fractional calculus and Prabhakar's theory. *Commun. Nonlinear Sci. Numer. Simul.* **2020**, *83*, 105114. [CrossRef]
30. Kochubei, A.N. General fractional calculus, evolution equations, and renewal processes. *Integral Equ. Oper. Theory* **2011**, *71*, 583–600. [CrossRef]
31. Brunner, H. *Volterra Integral Equations: An Introduction to Theory and Applications*; Cambridge University Press: Cambridge, UK, 2017; Volume 30.
32. Gripenberg, G.; Londen, S.O.; Staffans, O. *Volterra Integral and fUnctional Equations*; Cambridge University Press: Cambridge, UK, 1990; Volume 34.
33. Lubich, C. A stability analysis of convolution quadratures for Abel-Volterra integral equations. *IMA J. Numer. Anal.* **1986**, *6*, 87–101. [CrossRef]
34. Tsalyuk, Z. Volterra integral equations. *J. Sov. Math.* **1979**, *12*, 715–758. [CrossRef]
35. Doetsch, G. *Introduction to the Theory and Application of the Laplace Transformation*; Springer: Berlin/Heidelberg, Germany, 1974.
36. Horn, R.A.; Johnson, C.R. *Matrix Analysis*, 2nd ed.; Cambridge University Press: Cambridge, UK, 2013; p. xviii+643.
37. Garrappa, R.; Popolizio, M. Computing the matrix Mittag-Leffler function with applications to fractional calculus. *J. Sci. Comput.* **2018**, *77*, 129–153. [CrossRef]

Article

Handling Hysteresis in a Referral Marketing Campaign with Self-Information. Hints from Epidemics

Deborah Lacitignola

Department of Electrical and Information Engineering, University of Cassino and Southern Lazio, Di Biasio Str., I-03043 Cassino, Italy; d.lacitignola@unicas.it

Abstract: In this study we show that concept of backward bifurcation, borrowed from epidemics, can be fruitfully exploited to shed light on the mechanism underlying the occurrence of hysteresis in marketing and for the strategic planning of adequate tools for its control. We enrich the model introduced in (Gaurav et al., 2019) with the mechanism of self-information that accounts for information about the product performance basing on consumers' experience on the recent past. We obtain conditions for which the model exhibits a forward or a backward phenomenology and evaluate the impact of self-information on both these scenarios. Our analysis suggests that, even if hysteretic dynamics in referral campaigns is intimately linked to the mechanism of referrals, an adequate level of self-information and a fairly high level of customer-satisfaction can act as strategic tools to manage hysteresis and allow the campaign to spread in more controllable conditions.

Keywords: epidemic models; backward bifurcation; hysteresis; referral marketing; self-information

1. Introduction and Motivations

In the last century, mathematical models based on differential equations have been fruitfully applied to describe phenomena belonging to even extremely different disciplinary fields. As well known in literature [1], mathematical models can act essentially in two directions: those based on more sophisticated mathematical tools can give a great contribution in terms of quantitative predictions but simpler qualitative models can be precious to shed light on the constitutive mechanisms, highlighting their role and reciprocal interactions.

Precisely because they go to the heart of the phenomena, simple mechanistic qualitative models are capable to create bridges between apparently very distant worlds, making sure that models and methodologies used in a certain context could be exploited to open the way to the understanding of phenomena that are similar in their underlying mechanisms. On this line, it is not surprising that the simple mechanistic model found by Volterra [2] to describe the interaction between preys and predators in the Italian Adriatic Sea displays the same mathematical structure as the one introduced in those same years by Lotka [3] in the context of the chemical kinetics. And, again, it is not surprising at all that a discipline such as marketing has been able to benefit from the models and the modus operandi of mathematical epidemiology. In this case, the unifying factor is the idea of contagion, a key mechanism for those forms of marketing defined as viral. The viral name refers in particular to all those marketer-initiated consumer activities that spreads a marketing message unaltered across a market or segment in a limited time period mimicking an epidemic [4]. Terms from epidemiology have been hence widely used to explain such viral marketing process [5,6].

This interconnection has become even more pronounced with the unchallenged emergence of new means of communication. With consumers showing increasing resistance to traditional forms of advertising, marketers have been forced to rely on alternative strategies. Among these are social networks, whose usage is sensitively growing among marketing managers with the aim to promote an idea, a product or a brand at no additional cost to the firm. If a marketer encourages consumers to share and spread a marketing message through their social contacts, this is called Referral Marketing [7]. In few words, referral marketing spreads the word about a product or service through a business' existing customers, rather than traditional advertising. This kind of marketing uses referrals or word-of-mouth to promote services or products and businesses may control it through suitable strategies and make a viral referral campaign. Strategic use of referral marketing can hence allow marketers to leverage the power of consumer recommendations in order to achieve the desired results. On this line, questions as 'Which are the underlying set of interactions that ensure a marketing message to go viral? Which parameters can allow an effective spread of a marketing message through a viral process?' becomes simply crucial. And since a viral marketing message involves a person-to-person transmission spreading within a population just like an epidemic, it is not strange that the most likely enlightening answers could hence come from epidemic models.

In the classical models of epidemiology, the interactions between susceptible and infected is a key factor for the spread of an epidemic, qualitatively defined as a situation in which the number of the infected reaches a significant percentage at steady state. In the case of a viral marketing campaign it can be thought as a situation for which, because of the sharing mechanisms, the marketing message reaches and attracts a majority of its target consumers. Obviously in epidemiology one aims to contain epidemics whereas, within the marketing framework, the main purpose is to maximize the spread.

In the context of online social networks and digital contagion, many efforts have recently been made to model such kind of dynamics: reference [4] discussed the viral marketing diffusion within the SIR and SEIAR epidemic framework and [8] proposed a mathematical model borrowed from epidemiology to describe its spread. An extensive survey in reference [9] underlined that along with the viral component, a particular focus on customer behaviors should be given to ensure the relevance and survival of a newly-launched campaign. On this line, references [7,10] considered more realistic models for the viral campaign spread, where specific behavioral factors were introduced to take into account a customer's perspective about marketing messages, i.e., inherent adversion, brand trust, remembering and reminding.

In this paper we want to pursue this line focusing on the interplay between two behavioral mechanisms that can be involved together in a referral campaign. In fact, if referral is obviously the key mechanism of a referral campaign, it is not the only one. The Nielson Global Survey of Trust in Advertising [11] clearly supports the remarkable potential of referrals showing that for the question 'To what extent do you trust the following form of advertising?', the answer 'Recommendation from the people I know' gains the first position with 83%. But the answer 'Consumer's opinion posted online' is also on the podium with 66%, confirming how online reviews remain a trust source of customer information. This means that, on average, two-thirds of consumers feels the need of 'self-information' and make purchases after inspecting customers' opinions posted online about a particular product or service. In this case information comes from sources of reviews with no conflicts of interest, such as specific consumers' forum that collect opinions by those who bought particular products or experienced certain services.

Therefore in a referral marketing campaign, the nature of the information for the potential consumer can be twofold: passive, when it is linked to the mechanism of recommendations by friends and acquaintances or active when it is linked to the self-information mechanism described above. Our aim is to elucidate under what conditions the interplay between 'passive' and 'active' information can strengthen or weaken the survival chances of a referral campaign. On this line, we enrich the model introduced in [10] with the mechanism of self-information that accounts for information about the product performance basing on consumer's experience on the recent past. Such a mechanism, based on a kind of learning that a potential consumer can adopt during the referral campaign, is mathematically obtained by introducing a distributed lag in the population equations that therefore become an integro-differential system, i.e., a delay differential model. The importance of considering such kind of models is provided by the fact that the role of delays in biological [12–15] as well as in economic models [16–20] is widely recognized, being often appropriate for these kind of problems to allow the rate of change of the system variables to depend in some sense on the previous history. We want to establish conditions for which the model exhibits a forward or a backward phenomenology and evaluate the impact of self-information on both these scenarios. The backward phenomenology, in particular, is connected to a situation of bistability between the campaign-free equilibrium and the campaign-standing equilibrium and can lead the system towards hysteresis-type behaviors. In a very qualitative way the term hysteresis, related to the idea of "irreversibility", denotes the effects that persist after the causes that determined them have been removed. The relevance of using hysteresis at economic systems level is well recognized and marketing provides a generous framework to improve the understanding of this phenomenon in the economic sphere. In marketing, hysteresis is mainly thought in relation to consumer behaviour as well as to temporary or permanent changes of consumption patterns caused by specific marketing tools [21–23].

Its link with hysteresis is the reason why, in mathematical epidemiology, many papers have been focused on backward bifurcation, i.e., [24–28]. In that context, the basic reproduction number R_0 is usually defined as the expected number of new infections produced by a single infective individual introduced into a disease-free population [29] and $R_0 = 1$ represents the threshold value that separates the stability and instability regimes of the disease-free equilibrium. There are two bifurcation scenarios commonly detectable at $R_0 = 1$: (i) forward bifurcation that implies disease eradication below the threshold $R_0 = 1$; (ii) backward bifurcation that includes a saddle-node (sn) bifurcation at $R_0 = R_0^{sn} < 1$ along with a subcritical transcritical bifurcation at $R_0 = 1$; it involves a multiplicity of endemic equilibria and subcritical persistence of the disease. When a backward scenario is found, reducing R_0 below 1 is not sufficient to eradicate the disease and a further effort should be done until R_0 is lowered below the critical value R_0^{sn}. It is therefore obvious that, in epidemic models, detecting and managing the occurrence of backward bifurcations are two features of primary importance in the perspectives of the disease control. In viral marketing, however, the backward scenario may play a different role than in epidemics since it could be seen as an opportunity for the firm to carry on the viral campaign even in adverse conditions, which in itself adds an interesting perspective to the problem. Also in this case, however, the backward scenario must still be carefully monitored because in the bistability regime, too large displacements from the campaign-standing equilibrium can bring the system into the basin of attraction of the campaign-free equilibrium. That means a sudden collapse of the referral campaign.

The paper is structured as follows: In Section 2, we enrich the model introduced in [10] with the mechanism of self-information by the means of a variable that summarizes information about the product performance basing on consumers' experience on the recent past. In Section 3 we get conditions for the existence of a campaign-free and of a campaign-standing equilibria and establish under which conditions, expressed as a function of the system parameters, the campaign spread goes towards stopping. In Section 4 a bifurcation analysis in the neighbouring of the campaign-free equilibrium is performed and conditions are obtained for the emergence of a forward or a backward scenario that are also discussed in the perspective to improve the sustainability of the referral campaign. The effects of self-information on the bifurcation thresholds is shown in Section 5 where the role of the customer satisfaction parameter is also elucidated. Concluding remarks, in Section 6, close the paper.

2. A Referral Marketing Model with Self-Information

To mimics referral dynamics, a model was introduced in [7] with the total population divided in three mutually exclusive subpopulations: Unaware, Broadcaster and Inert. The unaware class U is the target market, namely 'susceptible' people that have not yet received the message about a certain product but are exposed and have a chance of receiving it; the broadcaster class B is composed of individuals who have received the message earlier and have the potential to spread the message further to their social contacts; the inert class I is instead made of individuals that, willingly or unwillingly, do not take part in the campaign even if they have come across it at least once. This model is essentially based on contagion as the basic transition mechanism between different subgroups.

To increase the degree of realism, the authors then proposed in [10] a more realistic model including some additional features raised in a survey campaign developed in [9]. Analyzing the surveys and the interactions between different people, they modeled the transition between different sub-groups with taking into account some additional factors that more clearly reflect customer's perspective about marketing messages: (a) inherent aversion, i.e., a portion of individuals could be strongly against the mechanisms of referral marketing in general; (b) brand trust, i.e., people need to 'trust' the person who is referring the product (for example family or friends) as well as the brand-names while participating in referral marketing; (c) remembering and reminding, i.e., strategically designed emails from the company or casual reminders from friends can tempt inert individuals to become broadcasters again. The following model was hence considered [10]:

$$\begin{aligned}
\dot{u} &= \mu - \rho b u - \mu u \\
\dot{b} &= p \rho b u - \sigma b + \alpha_1 b i - \mu b + \lambda i \\
\dot{i} &= (1-p) \rho b u + \sigma b - \alpha_1 b i - \lambda i - \mu i
\end{aligned} \quad (1)$$

where u, b and i are the fraction of the unaware, broadcaster and inert classed normalized by the total population. In (1), it is assumed that a broadcaster spreads the message to a member from unaware class at a rate ρ and, whenever a broadcaster sends the referral message to an unaware individual, this moves to the broadcaster class with a probability p and to the inert class with a probability $(1-p)$. The parameter $p \in [0,1]$ assumes a high value if the campaign comes from a trusted brand or the message comes from a trusted member and can be hence interpreted as the 'trust' parameter. The term $(1-p)$ accounts that some individuals of the unaware class might decide to ignore the messages or to not take part in the campaign, i.e., groups of individuals that are for example rigidly inert. Messages from not so trustworthy brand or members increases the value of $(1-p)$.

Once the unawares have become broadcasters or inert, they can 'change their mind' by moving from one class to another respectively. In fact, broadcasters can stop sharing the message, hence moving to the inert class at a rate σ. On the other hand, inert people can move back to broadcaster class following two different mechanisms: (i) independently of their interaction with other individuals (like reminder from the company etc.) at a rate λ or (ii) because of their interaction with another broadcaster (like reminder from a friend, discussion with family members) at a rate $\alpha_1 = \alpha p$ where α is the original relapse rate and p is the trust parameter. Obviously people can join or leave a particular social platform where the campaign is going. It is then assumed a constant input μ in the unaware class and a natural 'mortality rate' μ for each class so that a fixed population size can be maintained.

The analysis carried out in [10] showed that the brand loyalty and brand name are two important factors to create positive reaction of a person towards a campaign message. Moreover, model dynamics turned out to be critically affected by variations in the relapse rate α that was recognized to be crucial to safeguard the survival of the campaign. In particular, sufficiently high values of the relapse rate α could drive the system towards a bistability situation between the campaign-free and the campaign-standing equilibria.

In [10] the involved information mechanism was essentially passive because the spread of the message is based on referrals. To investigate the role of an active information on the spreading of the referral campaign, we equipped model (1) with a self-information variable m that summarizes information about the product performance basing on the customers' experiences in the recent past, i.e., online customer reviews. Because of this 'active' information process, we assume that unaware individuals can exit their class at a rate γ, moving to the broadcaster class with a probability q and to the inert class with a probability $(1-q)$. The parameter $q \in [0,1]$ assumes a high value if the online reviews on the product indicates an overall high level of satisfaction and can be hence interpreted as a 'customer satisfaction' parameter. We hence consider the following model:

$$\begin{aligned} \dot{u} &= \mu - \rho b u - \mu u - \gamma m u \\ \dot{b} &= p \rho b u - \sigma b + \alpha_1 b i - \mu b + \lambda i + \gamma q m u \\ \dot{i} &= (1-p) \rho b u + \sigma b - \alpha_1 b i - \lambda i - \mu i + \gamma (1-q) m u \end{aligned} \qquad (2)$$

where the self-information variable m is given by

$$m(t) = \int_{-\infty}^{t} f(u(\tau), b(\tau), i(\tau)) \, K_a(t-\tau) d\tau \qquad (3)$$

The distributed lag (3) in the governing equations means that unaware, broadcaster and inert individuals at time t are affected by the state variables u, b, i at possibly all previous times $\tau \le t$ in a way prescribed by the function $f(u(\tau), b(\tau), i(\tau))$ and distributed in the past by the delay kernel $K_a(t-\tau)$ which is also called 'memory function'.

We assume here that the function $f(u(\tau), b(\tau), i(\tau)) = kb$ where k is a positive parameter. Among the possible types of delay kernels, we consider

$$K_a(t) = a e^{-at} \qquad (4)$$

which qualitatively represents a weak delay in the sense that the maximum (weighted) response of the growth rate is to current population density whereas past densities have exponentially decreasing influence. Such a kernel provides therefore a reasonable effect of short term memory.

With (4) as delay kernel and by applying the *linear chain trick* [30], the set of delay differential Equations (2) and (3) turns out to be equivalent to the following set of ordinary differential equations that will be hereafter the object of our investigations:

$$\begin{aligned} \dot{u} &= \mu - \rho b u - \mu u - \gamma m u \\ \dot{b} &= p \rho b u - \sigma b + \alpha_1 b i - \mu b + \lambda i + \gamma q m u \\ \dot{i} &= (1-p) \rho b u + \sigma b - \alpha_1 b i - \lambda i - \mu i + \gamma (1-q) m u \\ \dot{m} &= a k b - a m \end{aligned} \qquad (5)$$

with $\alpha_1 = \alpha p$.

In the next section, we get conditions for the existence of a campaign-free and of a campaign-standing equilibria and establish under which conditions, expressed as a function of the system parameters, the campaign goes viral or is forced to stop.

3. The Campaign-Free and the Campaign-Standing Equilibria

Model (5) always admits a *campaign-free* equilibrium $E_0 = (1,0,0,0)$ and, under suitable conditions on the system parameters can admit one or two *campaign-standing* equilibrium $E^* = (u^*, b^*, i^*, m^*)$ where:

$$u^* = \frac{\mu}{b^*(\gamma k + \rho) + \mu}, \quad i^* = \frac{b^*(\gamma k + \rho)\sigma + \gamma k \mu (1-q) + \mu[\sigma + \rho(1-p)]}{(b^*\gamma k + b^*\rho + \mu)(b^*\alpha p + \lambda + \mu)}, \quad m^* = k b^*, \quad (6)$$

and b^* is a positive solution of the following algebraic equation,

$$P_2 b^2 + P_1 b + P_0 = 0, \quad (7)$$

with

$$P_2 = \alpha p (\gamma k + \rho),$$

$$P_1 = -p(\gamma k + \rho - \mu)\alpha + (\gamma k + \rho)(\lambda + \mu + \sigma) = p(\gamma k + \rho - \mu)(\alpha_0 - \alpha), \quad (8)$$

$$P_0 = (\mu - \gamma k - \rho)\lambda + \mu^2 + \mu\sigma - \gamma k \mu q - \rho \mu p = \mu(\sigma - \sigma_c),$$

and

$$\alpha_0 = \frac{(\gamma k + \rho)(\lambda + \mu + \sigma)}{p(\gamma k + \rho - \mu)}, \quad \sigma_c = \frac{1}{\mu}[\lambda(\gamma k + \rho - \mu) + \gamma k q \mu + \mu(p\rho - \mu)]. \quad (9)$$

By (6), it follows that E^* is a positive equilibrium provided b^* is a positive solution of (7). Moreover being (7) a second order algebraic equation we observe that, for certain ranges of the parameter values, model (5) could admit a multiplicity of campaign-standing equilibria.

In the next we assume $\mu \leq p\rho$ so that the natural "mortality rate" for each class is considered slow with respect to the marketing process. Under this condition, both σ_c and α_0 are positive quantities. We now determine the conditions for which model (5) can admit feasible (i.e. positive) campaign-standing equilibria. To do that, we inspect the discriminant of the algebraic Equation (7), namely

$$\Delta = P_1^2 - 4 P_2 P_0 = p^2 (\gamma k + \rho - \mu)^2 (\alpha_0 - \alpha)^2 - 4 \alpha p (\gamma k + \rho) \mu (\sigma - \sigma_c) \quad (10)$$

and observe that, if $\sigma < \sigma_c$, then (10) is a positive quantity so that by the Descartes' rule of signs, the algebraic Equation (7) admits only one positive real solution. On the contrary, if $\sigma > \sigma_c$, then $\Delta < 0 \Leftrightarrow \alpha_1 < \alpha < \alpha_2$, where

$$\alpha_{1/2} = \alpha_0 + \frac{Q_1 \mp \sqrt{3\alpha_0^2 Q_0^2 + Q_1^2 + 4\alpha_0 Q_0 Q_1}}{2 Q_0} \quad (11)$$

and

$$Q_0 = p^2(\gamma k + \rho - \mu)^2, \quad Q_1 = 4 p (\gamma k + \rho) \mu (\sigma - \sigma_c).$$

For $\sigma > \sigma_c$, Q_1 is a positive quantity and it is also easy to prove that $\alpha_1 < \alpha_0 < \alpha_2$. We can hence conclude that: if $\alpha_1 < \alpha < \alpha_2$ then Equation (7) admits no real solutions; if $\alpha < \alpha_1$ then, by the Descartes' rule of signs, the algebraic Equation (7) admits two negative real solutions; if $\alpha > \alpha_2$ then it admits two positive real solutions.

The above results can be summarized in the following theorems:

Theorem 1. *Let $\mu \leq p\rho$ and $\sigma < \sigma_c$. Then model (5) admits the campaign-free equilibrium $E_0 = (1,0,0,0)$ and one positive campaign-standing equilibrium E^*.*

Theorem 2. *Let $\mu \leq p\rho$ and $\sigma > \sigma_c$. Then model (5) admits the campaign-free equilibrium $E_0 = (1,0,0,0)$ and (i) if $\alpha < \alpha_2$, none positive campaign-standing equilibrium exists; (ii) if $\alpha > \alpha_2$, two positive campaign-standing equilibria exist.*

As far as the local stability properties of the campaign-free equilibrium $E_0 = (1,0,0,0)$ are concerned, we observe that the Jacobian matrix of model (5) when evaluated at E_0, is given by

$$J(E_0) = \begin{pmatrix} -\mu & -\rho & 0 & -\gamma \\ 0 & p\rho - \mu - \sigma & \lambda & \gamma q \\ 0 & (1-p)\rho + \sigma & -\lambda - \mu & \gamma(1-q) \\ 0 & ak & 0 & -a \end{pmatrix},$$

and admits $\omega = -\mu$ as an eigenvalue. To reason about the sign of the other three eigenvalues, we introduce the following matrices:

$$A = \begin{pmatrix} p\rho - \mu - \sigma & \lambda & \gamma q \\ (1-p)\rho + \sigma & -\lambda - \mu & \gamma(1-q) \\ ak & 0 & -a \end{pmatrix},$$

$$A_1 = \begin{pmatrix} -\lambda - \mu & \gamma(1-q) \\ 0 & -a \end{pmatrix}, A_2 = \begin{pmatrix} (1-p)\rho + \sigma & \gamma(1-q) \\ ak & -a \end{pmatrix}, A_3 = \begin{pmatrix} (1-p)\rho + \sigma & -\lambda - \mu \\ ak & 0 \end{pmatrix}.$$

and recall that the remaining three eigenvalues of $J(E_0)$ have negative real part if and only if the following conditions holds:

$$\det(A) < 0; \quad tr(A) < 0; \quad \sum_{i=1}^{3} \det(A_i) > \det(A)/tr(A).$$

We get $\det(A) = \mu a (\sigma_c - \sigma)$ and $tr(A) = p\rho - a - \lambda - 2\mu - \sigma$ so that:

$$\det(A) < 0 \Leftrightarrow \sigma > \sigma_c, \quad tr(A) < 0 \Leftrightarrow a > a_c$$

where σ_c is given in (9) and $a_c = p\rho - \lambda - 2\mu - \sigma$. We also observe that

$$a_c > 0 \Leftrightarrow \sigma < \tilde{\sigma} = p\rho - \lambda - 3\mu$$

and

$$\tilde{\sigma} < \sigma_c \Leftrightarrow -\mu < \gamma k q + \frac{\lambda(\rho + \gamma k)}{\mu}$$

that is always verified. Therefore for $\sigma > \sigma_c > \tilde{\sigma}$, the threshold quantity a_c is negative so that $tr(A) < 0$ for every positive value of a. Moreover by straightforward algebra follows that, for $\sigma > \sigma_c$, inequality $\sum_{i=1}^{3} \det(A_i) > \det(A)/tr(A)$ is always verified. We are hence in the position to state the following theorem:

Theorem 3. Let $\mu \leq p\rho$. (i) If $\sigma < \sigma_c$ then the campaign-free equilibrium E_0 is unstable. (ii) If $\sigma > \sigma_c$ then the campaign-free equilibrium E_0 is locally asymptotically stable.

In the following section, we analyze in more details the *nature* of the transcritical bifurcation at $\sigma = \sigma_c$ and its impact on the sustainability of the referral campaign.

4. Sustaining the Campaign: Forward or Backward Scenario?

Within the epidemic framework, backward scenarios have been mainly detected by the means of specific bifurcation approaches [31] with the aim to establish the nature of the bifurcation at $R_0 = 1$. Once the backward scenario is detected, the subcritical persistence of the disease can be prevented by varying significant parameters in the system or by the means of error-based methods as the Z-type control approach [32–34].

In this section, we discuss the occurrence of the backward vs the forward phenomenology for model (5), by using the method proposed in [35] that provides simple and manageable conditions for monitoring both these scenarios.

As shown in the previous section, $\sigma = \sigma_c$ is a transcritical bifurcation threshold. We observe that all the coefficients in the equilibrium Equation (7) may be regarded as functions of the parameter σ. Moreover at $\sigma = \sigma_c$, $P_0(\sigma_c) = 0$ so that Equation (7) becomes

$$P_2(\sigma_c)b^2 + P_1(\sigma_c)b = 0$$

and admits the roots $b = 0$ and $b = -\frac{P_1(\sigma_c)}{P_2(\sigma_c)}$. The former is related to the campaign-free equilibrium and the latter corresponds to a positive campaign-standing equilibrium only if $P_1(\sigma_c)$ and $P_2(\sigma_c)$ have opposite signs. Therefore, in order to have a positive campaign-standing equilibrium, $P_1(\sigma_c) < 0$ must hold. By implicit differentiation of Equation (7) with respect to σ, one obtains:

$$(2P_2 b + P_1)\frac{db}{d\sigma} + \frac{dP_2}{d\sigma}b^2 + \frac{dP_1}{d\sigma}b + \frac{dP_0}{d\sigma} = 0.$$

Now, looking at the equilibrium $b = 0$, at $\sigma = \sigma_c$ one has:

$$P_1(\sigma_c)\frac{db}{d\sigma}(\sigma_c) = -\frac{dP_0}{d\sigma} < 0, \qquad (12)$$

since, recalling (8), it holds $\frac{dP_0}{d\sigma} = \mu > 0$. Therefore, in order inequality (12) to be verified, $P_1(\sigma_c)$ and $\frac{db}{d\sigma}(\sigma_c)$ must have opposite sign. This means that the slope of the bifurcation curve at $b = 0$ must have opposite sign with respect to the coefficient $P_1(\sigma_c)$. Since in our case a forward scenario at $\sigma = \sigma_c$ is obtained when $\frac{db}{d\sigma}(\sigma_c) < 0$ and a backward scenario when $\frac{db}{d\sigma}(\sigma_c) > 0$, it hence follows that: (i) if $P_1(\sigma_c) < 0$ then a backward bifurcation occurs at $\sigma = \sigma_c$; (ii) if $P_1(\sigma_c) > 0$, the system displays a forward bifurcation at $\sigma = \sigma_c$.

For model (5), $P_1(\sigma_c) < 0$ is hence a necessary and sufficient condition for the occurrence of the backward bifurcation at $\sigma = \sigma_c$. By (9), the threshold α_0 depends on the parameter σ. Therefore by introducing,

$$\alpha^* = \alpha_0(\sigma_c) = \frac{\gamma k + \rho}{\gamma k + \rho - \mu}\tilde{\alpha}, \qquad (13)$$

where

$$\tilde{\alpha} = \frac{\gamma k(\mu q + \lambda) + \rho(\mu p + \lambda)}{\mu p}, \qquad (14)$$

the following result holds:

Theorem 4. *Let $\mu < p\rho$. (i) If $\alpha < \alpha^*$ then system (5) exhibits a forward bifurcation at $\sigma = \sigma_c$. (ii) If $\alpha > \alpha^*$ then system (5) exhibits a backward bifurcation at $\sigma = \sigma_c$.*

Proof. It follows from (8) by direct computations. □

Remark 1. *It easy to prove by direct computation that the threshold α_2 defined in (11) is such that*

$$\alpha_2 = \alpha_0(\sigma_c) = \alpha^*$$

so that results in Theorem 4 are in perfect agreement with the existence results provided in Theorem 2.

To validate the results found in Theorem 4, we show the local dynamics in the neighboring of the bifurcation value $\sigma = \sigma_c$ by the means of the bifurcation diagrams in the (σ, b^*) parameter space, Figure 1.

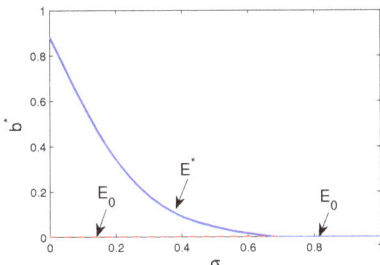 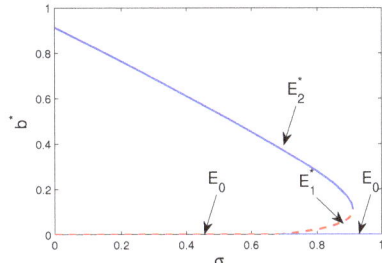

Figure 1. Bifurcation diagram in the plane (σ, b^*). The other parameters are $\mu = 0.05$; $\rho = 0.25$; $\lambda = 0.02$; $p = 0.7$; $q = 0.8$; $a = 0.5$; $k = 2$; $\gamma = 0.2$ so that $\alpha^* = 1.1684$ and $\sigma_c = 0.6850$. The solid lines (-) denote stability; the dashed lines (- -) denote instability. (**Left**) Forward scenario. The case $\alpha < \alpha^*$, $\alpha = 0.4$ – At $\sigma = \sigma_c = 0.6850$, system (5) exhibits a forward bifurcation. (**Right**) Backward scenario. The case $\alpha > \alpha^*$, $\alpha = 2$ – At $\sigma = \sigma_c = 0.6850$, system (5) exhibits a backward bifurcation. The value $\sigma_{SN} = 0.9105$ is the saddle-node bifurcation threshold.

For the numerical investigations, we decide to use the same parameters considered in [7,10] where a mathematical model was introduced basing on data collected through an extensive questionnaire-based-survey [9]. That survey recognized the dynamics of viral marketing propagation as a complicated nonlinear phenomenon that involves several interactions between the participants and is influenced by several intensive and extensive parameters. In [7,10], the above conceptual framework was developed through a mathematical ODE epidemic model that in [7] contains only the essential features of the phenomenon and in [10] is instead enriched with more realistic behavioral factors. The set of parameters used in these papers are chosen with the purposes (i) to illustrate the range of possible dynamics that can be expressed by the model and (ii) to elucidate which parameters and hence mechanisms can influence the overall dynamics. The perspective in which they move is a qualitative one and the model we develop in the present paper, enriching [10] with the self-information mechanism, moves exactly in the same qualitative direction. Therefore, to better elucidate the role of self-information and for a better comparison with the dynamics presented in [7,10], in the present study we have intentionally decided to consider the same set of parameters used there, namely: $\mu = 0.05$; $\rho = 0.25$; $\lambda = 0.02$; $p = 0.7$. The parameters for the self-information mechanism are instead chosen so that the hypothesis of Theorem 4 could be verified. We hence fix $q = 0.8$, $a = 0.5$, $k = 2$, $\gamma = 0.2$.

With this choice for the parameters, the assumption $\mu = 0.05 < (p\rho) = 0.1750$ is verified. Moreover $\alpha^* = 1.1684$ and $\sigma_c = 0.6850$.

In Figure 1 (left), the parameters are taken in order to verify condition (i) in Theorem 4 so that a forward scenario is obtained. In this case, $\alpha = 0.4 < \alpha^* = 1.1684$: a forward bifurcation occurs at $\sigma = \sigma_c = 0.6850$. For $\sigma < \sigma_c$, the campaign-standing equilibrium E^* is the only attractor for the system, being E_0 unstable in this range. Differently, for $\sigma > \sigma_c$, the campaign-free equilibrium E_0 is the only attractor for the system and increasing σ above the threshold σ_c is sufficient to stop the campaign. In Figure 1 (right), we choose the parameter values so that condition (ii) in Theorem 4 is verified. In this case, $\alpha = 2 > \alpha^* = 1.1684$: a backward bifurcation occurs at $\sigma = \sigma_c = 0.6850$ and $\sigma_{SN} = 0.91058$ is the saddle node-bifurcation value. For $\sigma < \sigma_c$, the campaign-standing equilibrium E^* is the only attractor for the system since E_0 is unstable in this range. For $\sigma_c < \sigma < \sigma_{SN}$, a bistability situation occurs, with the disease-free equilibrium E_0 and the endemic equilibrium E_2 as local attractors. For $\sigma > \sigma_{SN}$, the campaign-free equilibrium E_0 becomes the only attractor for the system. In this case, the value of the parameter σ should be increased above the saddle-node bifurcation threshold σ_{SN} in order to stop the campaign.

The above results well put into evidence that the sustainability of the referral campaign is linked to the suitable interplay between the two parameters α and σ that respectively regulate the reciprocal transition between the broadcaster and inert classes. We recall that α is the relapse rate from the inert to the broadcaster class whereas σ is the dropout rate of the broadcaster class in favor of the inert class. Therefore, when the impact of the relapse rate α is below a certain threshold, i.e., $\alpha < \alpha^*$, then increasing the dropout rate σ above a certain threshold σ_c has the effect to stop the campaign. On the contrary, when the impact of the relapse rate α is much stronger, i.e., $\alpha > \alpha^*$, then simply increasing the dropout rate σ above σ_c is not enough to stop the campaign and the value of σ must exceed an higher threshold σ_{SN} to make it end. This aspect would seem to suggest that a backward scenario could strengthen the campaign's chances of survival. However, in the bistability range $\sigma_c < \sigma < \sigma_{SN}$, the dynamics of the system is highly dependent on the initial conditions so that, within the backward scenario, a sudden stop of the campaign could likely occur.

In this latter case, inducing a slight reduction of the dropout rate σ does not allow to restore the spreading of the campaign. To this aim, it is in fact necessary to drastically reduce σ below the σ_c value. This behavior is depicted in Figure 2 and clearly indicates a hysteretic phenomenology since the functioning and the current state of the system can be understood in a more detailed manner with reference to its past. In this sense, the effects on the dynamics persist after the causes that determined them have been removed.

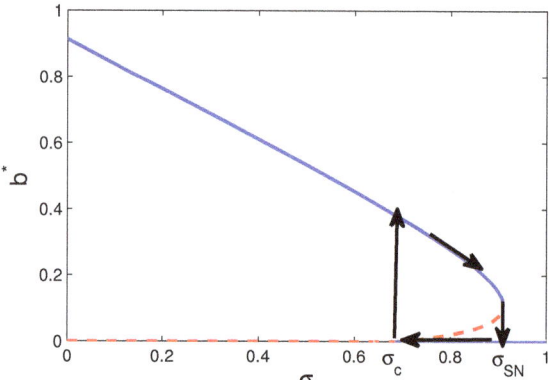

Figure 2. Graphical representation of an hysteresis cycle on the bifurcation diagram in the plane (σ, b^*) in the case $\alpha > \alpha^*$, where a backward scenario is obtained. The other parameters are as in Figure 1 (right). Here $\alpha^* = 1.1684$, $\sigma_c = 0.6850$ and the value $\sigma_{SN} = 0.9105$ is the saddle-node bifurcation threshold. The solid lines (-) denote stability; the dashed lines (- -) denote instability.

As a consequence if the backward phenomenology can represent an opportunity, it nevertheless introduces a risk factor and, for this reason, it must be detected and adequately managed. This suggests the need for a more accurate characterization of the bistability range delimited by the transcritical threshold σ_c and by the saddle-node threshold σ_{SN}. To this aim, we derive the analytical expression of the saddle-node bifurcation threshold σ_{SN}. We first recall that the two campaign-standing equilibria E_1^* and E_2^* are such that:

$$E_1 = (u_1^*, b_1^*, i_1^*, m_1^*), \quad E_2 = (u_2^*, b_2^*, i_2^*, m_2^*),$$

where u_i, i_i and m_i are defined in (6) and b_i are the two positive solutions of the algebraic Equation (7) whose coefficients are defined in (8). More precisely,

$$b_{1/2} = -\frac{P_1 \mp \sqrt{\Delta}}{2 P_2}$$

with Δ defined in (10) and the quantities α_0 and σ_c defined in (9). At $\sigma = \sigma_{SN}$, the two campaign-standing equilibria E_1^* (unstable) and E_2^* (stable) coalesce and disappear so that, for $\sigma > \sigma_{SN}$ the campaign-free equilibria is the only attractor for the system. The saddle-node bifurcation of the two campaign-standing equilibria can be detected by requiring that $\Delta = 0$ so that $E_1^* \equiv E_2^*$. At this regard, it holds:

$$\Delta = 0 \Leftrightarrow \sigma_{1/2} = \frac{1}{(\gamma k + \rho)} \left[\alpha p (\gamma k + \mu + \rho) - (\gamma k + \rho)(\lambda + \mu) \mp 2\sqrt{\Delta^*} \right]$$

where

$$\Delta^* = \alpha p (\gamma k + \rho) \mu p (\alpha - \tilde{\alpha}) \qquad (15)$$

with $\tilde{\alpha}$ defined in (14). By direct computation it easy follows that if $\alpha > \tilde{\alpha}$ then σ_i are real quantities and that, for $\alpha > \alpha^* > \tilde{\alpha}$, the inequalities $\sigma_1 > \sigma_c > 0$ hold. Therefore,

$$\sigma_{SN} = \sigma_1 = \frac{1}{(\gamma k + \rho)} \left[\alpha p (\gamma k + \mu + \rho) - (\gamma k + \rho)(\lambda + \mu) - 2\sqrt{\Delta^*} \right]$$

is the saddle-node bifurcation threshold and $[\sigma_c, \sigma_{SN}]$ is the bistability range for model (5). In the next section, we show how these critical thresholds are affected by variations in the self-information level.

5. Effects of Self-Information on the Bifurcation Thresholds

Since γ and k are the parameters specifically related to the self-information mechanism, we introduce the *information parameter* $\zeta = \gamma k$ and consider the different bifurcation thresholds as function of ζ, i.e.,

$$\alpha^*(\zeta) = \frac{\zeta + \rho}{\zeta + \rho - \mu} \frac{\zeta(\mu q + \lambda) + \rho(\mu p + \lambda)}{\mu p}$$

$$\sigma_c(\zeta) = \frac{1}{\mu} [\lambda(\zeta + \rho - \mu) + \zeta q \mu + \mu(p\rho - \mu)] \qquad (16)$$

$$\sigma_{SN}(\zeta) = \frac{1}{(\zeta + \rho)} \left[\alpha p (\zeta + \mu + \rho) - (\zeta + \rho)(\lambda + \mu) - 2\sqrt{\Delta^*(\zeta)} \right]$$

with $\Delta^*(\zeta)$ as defined in (15). We observe that the saddle-node bifurcation threshold σ_{SN} is a real quantity provided that the information variable ζ is chosen in the range $(0, \zeta^*)$, where

$$\zeta^* = \frac{\alpha \mu p - \rho(\mu p + \lambda)}{\mu q + \lambda} \qquad (17)$$

Moreover, since in the backward scenario $\alpha > \alpha^*$, the inequality

$$\alpha > \frac{\rho(\mu p + \lambda)}{\mu p},$$

is always verified and ζ^* is a positive quantity. We also observe that the transcritical bifurcation threshold σ_c is an increasing function of ζ, being

$$\frac{d\sigma_c}{d\zeta} = \frac{\mu q + \lambda}{\mu}.$$

Therefore an higher information increases the threshold σ_c, so that both in the forward and in the backward regime it becomes larger the range $[0, \sigma_c]$ for which the campaign-standing equilibrium is the only attractor for the system.

Moreover, Figure 3 also indicates that:

- the threshold α^* increases with increasing the information variable ζ. This means that an higher information increases the threshold α^*, favoring the forward regime with respect to the backward scenario. In this sense, information would act as a stabilizing mechanism;
- within the backward scenario, the saddle-node bifurcation threshold σ_{SN} increases with increasing the information variable ζ. This means that an higher information implies a higher value of σ in order to stop the campaign. However, the length of the bistability range $[\sigma_c, \sigma_{SN}]$ does not have a monotone trend as function of the information variable ζ. More precisely, for intermediate values of ζ, the bistability range decreases whereas it increases when the values of ζ are too small or too large. This would qualitatively mean that too much or too little information, although enlarging the chances of survival of the campaign, can have eventually a destabilizing effect on the system dynamics favoring sudden collapses in broadcasters that could lead to a sudden stop of the campaign according to a hysteretic phenomenology.

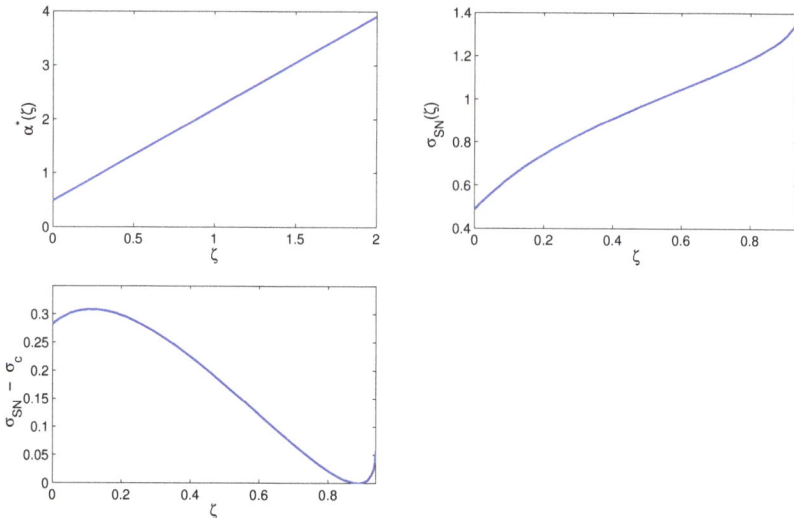

Figure 3. Thresholds (16) as function of the information variable ζ. The other parameters are chosen as in Figure 1. (**Top-left**) The threshold $\alpha^*(\zeta)$ as function of ζ. (**Top-right**) The saddle-node bifurcation threshold $\sigma_{SN}(\zeta)$ as function of ζ. The threshold σ_{SN} is feasible in the range $(0, \zeta^*]$, with $\zeta^* = 0.9375$ (**Bottom**) The length of the bistability range, i.e., $\sigma_{SN} - \sigma_c$, within the backward scenario as function of ζ. The bistability range is increasing for $[0, \zeta_1)$ and (ζ_2, ζ^*) and it decreases for $[\zeta_1, \zeta_2]$. Here $\zeta^* = 0.9375$; $\zeta_1 = 0.1135$; $\zeta_2 = 0.8861$.

To give a more quantitative measure of the impact of the information parameter ζ on the bifurcation thresholds (16), we will make use of the sensitivity analysis that is a useful tool to reveal how a certain parameter can influence the campaign transmission. The sensitivity of a certain variable with respect to system parameters can be measured through a sensitivity index that provides a quantitative measure of the relative change in a variable when a parameter changes. When the variable is a differentiable function of the parameter, the sensitivity index is defined as follows:

Definition 1. [36] *The normalized forward sensitivity index of a variable u, that depends differentiably on a parameter p, is defined as*

$$\phi_p^u = \frac{\partial u}{\partial p} \frac{p}{u}$$

The normalized forward sensitivity index of a variable with respect to a parameter is therefore the ratio of the relative change in the variable to the relative change in the parameter.

Figure 4 shows how the sensitivity index of the different thresholds α^*, σ_c and σ^{SN} varies with varying the information parameter ζ. For both α^* and σ_c, the sensitivity index is a saturating function of ζ, the first increasing more slowly than the latter. The sensitivity of σ^{SN} instead rapidly grows for enough low values and for enough high values of the information parameter ζ; on the contrary, it grows very slowly for intermediate values of ζ. In Table 1, we show more quantitatively how variations in the information parameter ζ can affect the different thresholds (16).

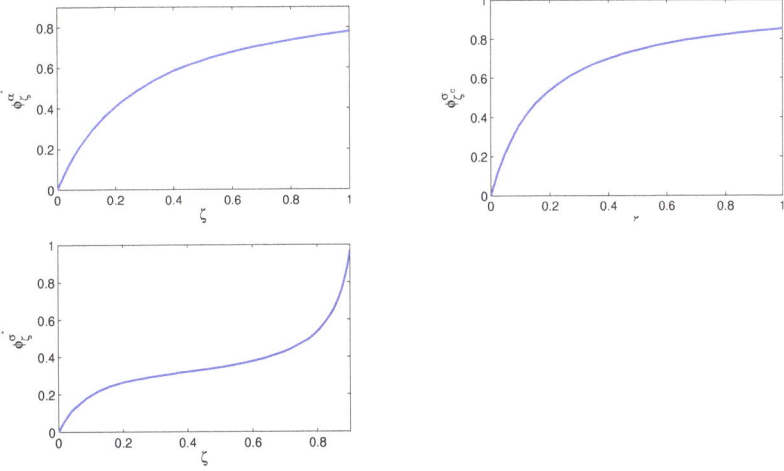

Figure 4. Sensitivity indices of the different thresholds α^*, σ_c and σ^{SN} as function of the information variable ζ. The other parameters are chosen as in Figure 1. (**Top-left**) Plot of the sensitivity $\phi_\zeta^{\alpha^*}$ versus the information variable ζ; (**Top-right**) Plot of the sensitivity $\phi_\zeta^{\sigma_c}$ versus the information variable ζ; (**Bottom**) Plot of the sensitivity $\phi_\zeta^{\sigma^{SN}}$ versus the information variable ζ.

Table 1. Sensitivity indices of the thresholds α^*, σ_c and σ^{SN} for three different levels of information: low, intermediate and high. The numerical values of the system parameters used for the computations are: $\mu = 0.05$; $\rho = 0.25$; $\lambda = 0.02$; $p = 0.7$; $q = 0.8$. Here $\zeta^* = 0.9375$; $\zeta_1 = 0.1135$; $\zeta_2 = 0.8861$.

Low Information $0 < \zeta < \zeta_1$	Intermediate Information $\zeta_1 < \zeta < \zeta_2$	High Information $\zeta_2 < \zeta < \zeta^*$
$\zeta = 0.08$	$\zeta = 0.5$	$\zeta = 0.9$
$\phi_\zeta^{\alpha^*} = 0.2154$	$\phi_\zeta^{\alpha^*} = 0.6380$	$\phi_\zeta^{\alpha^*} = 0.7614$
$\phi_\zeta^{\sigma_c} = 0.38$	$\phi_\zeta^{\sigma_c} = 0.74$	$\phi_\zeta^{\sigma_c} = 0.8404$
$\phi_\zeta^{\sigma_{SN}} = 0.1734$	$\phi_\zeta^{\sigma_{SN}} = 0.3450$	$\phi_\zeta^{\sigma_{SN}} = 0.9713$

It is interesting to observe that ζ affects such thresholds differently depending on the level of information we consider.

In the case of low information, σ_c is the most affected threshold: in fact, $\phi_\zeta^{\sigma_c} = 0.38$, which means that increasing (or decreasing) the parameter ζ by 10%, increases (or decreases) the transcritical threshold σ_c by 3.8%. The less affected threshold is instead σ_{SN}, being $\phi_\zeta^{\sigma_{SN}} = 0.1734$. However, for this case, the sensitivity indices for the three thresholds have numerical values fairly low and quite similar each others. A similar situation, but with higher values of the sensitivity indices is found for the case of intermediate levels of information for which the thresholds α^* and σ_c are influenced by variations in the information parameter ζ much more than the saddle-node bifurcation threshold σ_{SN}. Also in this case, σ_c is the threshold most influenced by variations in the information parameter, being $\phi_\zeta^{\sigma_c} = 0.74$; the less affected threshold is instead σ_{SN}, since $\phi_\zeta^{\sigma_{SN}} = 0.3450$. The case of high levels of information presents a completely different scenario being now σ_{SN} the most affected threshold with $\phi_\zeta^{\sigma_{SN}} = 0.97$: this means that increasing (or decreasing) the parameter ζ by 10%, increases (or decreases) the saddle-node bifurcation threshold σ_{SN} by 9.7%. In this case, however, also the thresholds α^* and σ_c are significantly influenced by variations in the information parameter ζ.

These results seem to suggest that *intermediate* levels of information allow to spread the campaign in more controllable conditions. In fact, they seem to (i) favor a forward-type regime over a backward type, as it can be observed by the significant increase in the α^* threshold; (ii) favor the presence of a single campaign-standing type attractor (significant increase in the threshold σ_c) with respect to a bistability regime (loose impact on the threshold σ_{SN}). In this sense, intermediate levels of information are surely preferable to low ones. On the other hand, too high levels of information sensitively impact the saddle-node threshold, favoring a bistability situation where the chances of the campaign's survival increase despite being exposed to the likely emergence of hysteretic dynamics.

The survival of the campaign obviously depends on the number of people who make it to spread and in the bistability range, when σ tends to σ_{SN}, the level of broadcasters at the campaign-standing equilibrium tends to decrease, as it can be seen from the bifurcation diagram in Figure 1.

It is therefore interesting to ask whether the level of customer satisfaction linked to the self-information process can act as a destabilizing factor for the survival of the campaign. Numerical simulations in Figure 5 (Top) show that, for low or intermediate values of the self-information parameter ζ, the campaign-standing equilibrium is rather resilient to variations in the level of the customer satisfaction q. However, increasing the level of self-information from low to intermediate, the impact of q also increases to the point that, for high values of ζ, a threshold value q^* can be found below which the referral campaign is driven to stop, Figure 5 (Bottom).

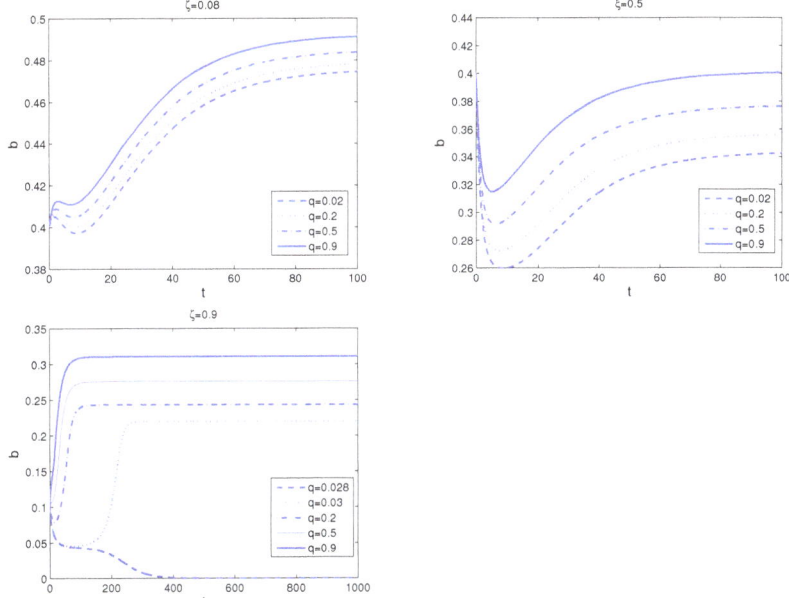

Figure 5. Impact of the customer satisfaction parameter q on the referral campaign in the bistability region, $\sigma \in [\sigma_c, \sigma_{SN}]$, for different levels of self-information. Initial conditions are chosen in the neighbouring of the campaign-standing equilibrium. The other parameters are as in Figure 1. (**Top-left**) Low level of the self-information parameter, i.e., $\zeta = 0.08$ ($k = 2; \gamma = 0.04$) and $\sigma = 0.4$. (**Top-right**) Intermediate level of the self-information parameter, i.e., $\zeta = 0.5$ ($k = 2; \gamma = 0.25$) and $\sigma = 0.7$. (**Bottom**) High level of the self-information parameter, i.e., $\zeta = 0.9$ ($k = 2; \gamma = 0.45$) and $\sigma = 0.9$.

6. Conclusions

With this study we wanted to show that the concept of backward bifurcation, borrowed from epidemics, can be fruitfully exploited to shed light on the mechanism underlying the occurrence of hysteresis in marketing as well as for the strategic planning of adequate tools for its control.

In this paper, we considered a referral marketing model with self-information and evaluated how the interplay between a passive information (due to referrals) and an active information (due to self-information) impacts the sustainability of the viral campaign. We found that the emergence of a forward or a backward phenomenology is essentially linked to passive information mechanisms since the occurrence of these scenarios depends on the suitable interplay between the two parameters that regulate the reciprocal transition between the broadcaster and inert classes by the means of referrals.

Differently from epidemics, in the viral marketing context, a backward scenario could strengthen the campaign's chances of survival. But if it can represent an opportunity from one side, on the other it introduces a risk factor because of the bistability range where system dynamics highly depends on the initial conditions. In this range hysteresis-type behaviors can hence emerge. Moreover, if in epidemics the main purpose is to 'avoid' a backward type scenario, for viral marketing this aim becomes learning to tame and eventually manage the backward phenomenology. In the present study, this has been shown to be the role of self-information that, however, needs to be properly calibrated. According to the Latin sentence 'in medio stat virtus', our analysis shows in fact that intermediate levels of self-information allow the campaign to spread in more controllable conditions by favoring the more reassuring forward-type regime over the backward one and, in both these cases, by widening the range of parameters in which the campaign-standing equilibrium is the only attractor for the system. Too high levels of information can instead broaden the region of parameters in which bistability occurs and, although enlarging the chances of survival of the campaign, can be responsible of sudden collapses in its spread. Just in this case, the level of customer satisfaction turns out to have a certain weight since a threshold customer satisfaction value can be found, below which, small fluctuations from the campaign-standing equilibrium value can lead the campaign to a sudden stop.

Therefore, even if hysteretic dynamics in referral campaigns may likely occur because intimately linked to the mechanism of referrals, an adequate level of self-information and a fairly high level of customer-satisfaction, can be two weapons capable to control hysteresis by transforming a potential risk into an opportunity.

In conclusion, this study represents a qualitative step to better understand how self-information can impact the sustainability of a referral marketing campaign and, within such a qualitative dimension, there is no presumption to fit the trend of a specific campaign. To provide further insight into the topic, two extensions are currently the subject of ongoing research: (i) giving the model a more quantitative dimension through a validation with a practical experience and (ii) exploring the possible impact of multilayer or multiplex networks, that may lead to some hidden patterns of influence and interplay between the self-information mechanism and the viral spreading of the campaign.

Funding: This research received no external funding.

Institutional Review Board Statement: Not applicable.

Informed Consent Statement: Not applicable.

Acknowledgments: The present work has been performed under the auspices of the Italian National Group for Mathematical Physics (GNFM-Indam). The author wishes to thank the anonymous Referees and the handling Editor for their valuable comments and remarks.

Conflicts of Interest: The author declares no conflict of interest.

References

1. May, R.M. Uses and Abuses of Mathematics in Biology. *Science* **2004**, *303*, 790–793. [CrossRef] [PubMed]
2. Volterra, V. Variazioni e fluttuazioni del numero di individui in specie animali conviventi. *Mem. Acc. Lincei* **1926**, *2*, 31–113.
3. Lotka, A. *Elements of Physical Biology*; William and Wilkins: Baltimore, MD, USA, 1925.
4. Sohn, K.; Gardner, J.; Weaver, J. Viral Marketing–More than a Buzzword. *J. Appl. Bus. Econ.* **2013**, *14*, 21–42.
5. Kaplan, A.M.; Haenlein, M. Two hearts in three-quarter time: How to waltz the social media/viral marketing dance. *Bus. Horizons* **2011**, *54*, 253–263. [CrossRef]
6. Reichstein, T.; Brusch, I. The decision-making process in viral marketing—A review and suggestions for further research. *Psychol. Mark.* **2019**, *36*, 1062–1081. [CrossRef]
7. Bhattacharya, S.; Gaurav, K.; Ghosh, S. Viral marketing on social networks: An epidemiological perspective. *Phys. A Stat. Mech. Appl.* **2019**, *525*, 478–490. [CrossRef]
8. Rodrigues, H.; Fonseca, M. Can information be spread as a virus? viral marketing as epidemiological model. *Math. Methods Appl. Sci.* **2016**, *39*, 4780–4786. [CrossRef]
9. Ghosh, S.; Bhattacharya, S.; Gaurav, K.; Singh, Y. Going Viral: The Epidemiological Strategy of Referral Marketing. *arXiv* **2018**, arXiv:1808.03780.
10. Ghosh, S.; Gaurav, K.; Bhattacharya, S.; Singh, Y.N. Ensuring the Spread of Referral Marketing Campaigns: A Quantitative Treatment. *Sci. Rep.* **2020**, *10*, 11072.
11. The Nielsen Company. Global Trust in Advertising. 2015. Available online: https://www.nielsen.com/wp-content/uploads/sites/3/2019/04/global-trust-in-advertising-report-sept-2015-1.pdf (accessed on 14 March 2021).
12. Beretta, E.; Breda, D. Discrete or distributed delay? Effects on stability of population growth. *Math. Biosci. Eng.* **2016**, *13*, 19. [CrossRef]
13. Buonomo, B.; d'Onofrio, A.; Lacitignola, D. Global stability of an SIR epidemic model with information dependent vaccination. *Math. Biosci.* **2008**, *216*, 9–16. [CrossRef] [PubMed]
14. Feng, J.; Sevier, S.; Huang, B.; Jia, D.; Levine, H. Modeling delayed processes in biological systems. *Phys. Rev. E* **2016**, *94*, 032408. [CrossRef]
15. Rombouts, J.; Vandervelde, A.; Gelens, L. Delay models for the early embryonic cell cycle oscillator. *PLoS ONE* **2018**, *13*, e0194769. [CrossRef]
16. Bischi, I.; Naimzada, A. A Kaleckian Macromodel with Memory. In *Cycles, Growth and the Great Recession*; Cristini, A., Leoni, R., Eds.; Routledge: London, UK, 2015.

17. Matsumoto, A.; Chiarella, C.; Szidarovszky, F. Dynamic monopoly with bounded continuously distributed delay. *Chaos Solitons Fractals* **2013**, *47*, 66–72. [CrossRef]
18. Krawiec, A.; Szydłowski, M. Economic growth cycles driven by investment delay. *Econ. Model.* **2017**, *67*, 175–183. [CrossRef]
19. Sîrghi, N.; Neamtu, M.; Mircea, G.; Ramescu, D. The Deterministic Model With Time Delay for a New Product Diffusion in a Market. *Timis. J. Econ. Bus.* **2018**, *11*, 55–66. [CrossRef]
20. Hughes, C.; Swaminathan, V.; Brooks, G. Driving Brand Engagement Through Online Social Influencers: An Empirical Investigation of Sponsored Blogging Campaigns. *J. Mark.* **2019**, *83*, 78–96. [CrossRef]
21. Kryukov, E.; Malgin, V.; Malgina, I. The influence of Hysteresis in consumer's behaviour for premium price evaluation. *Manag. Mark. J.* **2014**, *12*, 205–218.
22. Hanssens, D. Keeps Working and Working and Working . . . The Long-Term Impact of Advertising. *NIM Mark. Intell. Rev.* **2015**, *7*, 42–47. [CrossRef]
23. Moraru, A.; Barbulescu, A.; Duhnea, C. Consumption and hysteresis: The new, the old, and the challenge. *Econ. Res. Ekon. Istraz.* **2018**, *31*, 1965–1980. [CrossRef]
24. Anguelov, R.; Garba, S.; Usaini, S. Backward bifurcation analysis of epidemiological model with partial immunity. *Comput. Math. Appl.* **2014**, *68*, 931–940. [CrossRef]
25. Buonomo, B.; Lacitignola, D. On the backward bifurcation of a vaccination model with nonlinear incidence. *Nonlinear Anal. Model. Control* **2011**, *16*, 30–46. [CrossRef]
26. Gumel, A. Causes of backward bifurcations in some epidemiological models. *J. Math. Anal. Appl.* **2012**, *395*, 355–365. [CrossRef]
27. Lacitignola, D.; Saccomandi, G. Managing awareness can avoid hysteresis in disease spread: An application to Coronavirus Covid-19. *Chaos Solitons Fractals* **2021**, *144*, 110739. [CrossRef] [PubMed]
28. Zhang, W.; Wahl, L.; Yu, P. Backward bifurcations, turning points and rich dynamics in simple disease models. *J. Math. Biol.* **2016**, *73*, 947–976. [CrossRef]
29. Van den Driessche, P.; Watmough, J. A simple SIS epidemic model with a backward bifurcation. *J. Math. Biol.* **2000**, *40*, 525–540. [CrossRef] [PubMed]
30. Smith, H. *An Introduction to Delay Differential Equations with Applications to the Life Sciences*; Springer Science & Business Media: Berlin/Heidelberg, Germany, 2010.
31. Strogatz, S. *Nonlinear Dynamics and Chaos: With Applications to Physics, Biology, Chemistry, and Engineering*, 2nd ed.; Studies in Nonlinearity; Westview Press: Nashville, TN, USA, 2014.
32. Guo, D.; Zhang, Y. Neural Dynamics and Newton–Raphson Iteration for Nonlinear Optimization. *J. Comput. Nonlinear Dyn.* **2014**, *9*, 021016. [CrossRef]
33. Lacitignola, D.; Diele, F. On the Z-type control of backward bifurcations in epidemic models. *Math. Biosci.* **2019**, *315*, 108215. [CrossRef] [PubMed]
34. Lacitignola, D.; Diele, F. Using awareness to Z-control a SEIR model with overexposure. Insights on Covid-19 pandemic. *Chaos Solitons Fractals* **2021**, under revision.
35. Brauer, F. Backward bifurcations in simple vaccination models. *J. Math. Anal. Appl.* **2004**, *298*, 418–431. [CrossRef]
36. Chitnis, N.; Hyman, J.; Cushing, J. Determining Important Parameters in the Spread of Malaria Through the Sensitivity Analysis of a Mathematical Model. *Bull. Math. Biol.* **2008**, *70*, 1272–1296. [CrossRef] [PubMed]

Article

Mass-Preserving Approximation of a Chemotaxis Multi-Domain Transmission Model for Microfluidic Chips

Elishan Christian Braun [1,†], Gabriella Bretti [2,*,†] and Roberto Natalini [2,†]

[1] Department Mathematics, University of Rome 3, 00146 Rome, Italy; Elishan@hotmail.de or elishanchristian.braun@uniroma3.it
[2] Istituto per le Applicazioni del Calcolo "M.Picone", 00185 Rome, Italy; r.natalini@iac.cnr.it
[*] Correspondence: g.bretti@iac.cnr.it
[†] These authors contributed equally to this work.

Abstract: The present work is inspired by the recent developments in laboratory experiments made on chips, where the culturing of multiple cell species was possible. The model is based on coupled reaction-diffusion-transport equations with chemotaxis and takes into account the interactions among cell populations and the possibility of drug administration for drug testing effects. Our effort is devoted to the development of a simulation tool that is able to reproduce the chemotactic movement and the interactions between different cell species (immune and cancer cells) living in a microfluidic chip environment. The main issues faced in this work are the introduction of mass-preserving and positivity-preserving conditions, involving the balancing of incoming and outgoing fluxes passing through interfaces between 2D and 1D domains of the chip and the development of mass-preserving and positivity preserving numerical conditions at the external boundaries and at the interfaces between 2D and 1D domains.

Keywords: multi-domain network; transmission conditions; finite difference schemes; chemotaxis; reaction-diffusion models

MSC: 65M06; 35L50; 92B05; 92C17; 92C42

Citation: Braun, E.C.; Bretti, G.; Natalini, R. Mass-Preserving Approximation of a Chemotaxis Multi-Domain Transmission Model for Microfluidic Chips. *Mathematics* 2021, 9, 688. https://doi.org/10.3390/math9060688

Academic Editor: Ioannis G. Stratis

Received: 3 March 2021
Accepted: 18 March 2021
Published: 23 March 2021

Publisher's Note: MDPI stays neutral with regard to jurisdictional claims in published maps and institutional affiliations.

Copyright: © 2021 by the authors. Licensee MDPI, Basel, Switzerland. This article is an open access article distributed under the terms and conditions of the Creative Commons Attribution (CC BY) license (https://creativecommons.org/licenses/by/4.0/).

1. Introduction

The aim of the present work is to study both the modelling and numerical approximation of a chemotaxis-reaction-diffusion mathematical system describing the qualitative behavior of different cell species living in a confined environment. This work is inspired by laboratory experiments made on microfluidic chip [1], where some populations coexist and interact. In recent years, there has been the development of a new approach to biological studies aimed at reconstructing organs and complex biological processes on a chip [2]. The fundamental idea is that the comprehension of the sophisticated physiology of organisms, based on the complex behavior and interaction of cell populations, tissues, and organs, needs interdisciplinary contributions from biology to mathematics.

Motivated by the laboratory setting of the experiment in microfluidic chips [1–3]—see also the short description of the experiments reported in Section 2.1.1—we introduce a model mimicking the interactions between two cell populations—namely, immune and cancer cells.

The mathematical model, proposed in Section 2.2, is a reaction-diffusion system with chemotaxis and describes birth and death processes, the migration of immune cells driven by chemical signals produced by tumor cells, and interactions between different cell species. We underline that, since the chemical gradients are not measured experimentally, by using the simulation algorithm based on the mathematical model proposed here, the chemical concentration gradients in the chip can be obtained by solving the inverse problem of minimizing the residuals between the measured trajectories and the simulated ones; see also the discussion in Section 5 about the future developments of our work. From a mathematical point of view, we follow the framework of the classical macroscopic models of chemotaxis; see, for instance, [4], where the evolution of the density of cells is described by a parabolic equation and the concentration of a chemoattractant can be given by a parabolic or elliptic equation, depending on the different regimes to be described and on the authors' choices. The choice of a continuous model to reproduce an experiment in a confined environment, with a relatively small number of cells, is motivated by the fact that we aim at developing a simulation tool which is able to describe the phenomena of immunosorveillance of cancer in tissues, where billions of cells are present. For this reason, a macroscopic model is more suitable respect a particle model.

In the chambers, we consider a 2D doubly parabolic model which is a modification of the Keller–Segel model [4] to take into account the presence of two populations both producing chemical signal which are interacting each other. We remark that we consider only the 2D case, since the experimental data do not take into account the height of the chip. Clearly, in principle, our framework (model and numerical algorithm) could be easily extended to the third dimension and we remark that, analogously to the 2D-1D case here considered, mass-preserving and positivity preserving numerical conditions at the external boundaries and at the interfaces between 3D and channels still hold.

We consider the microchannels connecting the 2D chambers as 1D lines for modelling and computational reasons, as explained in Section 1.1. In order to model the dynamics on the microchannels, we choose two different approaches: we can assign a 1D version of the doubly parabolic model used in the chambers; otherwise, we can assign a model derived from a 1D-GA model [5], being characterized by the more realistic feature that the speed of propagation of cells in the channels is finite, which seems the dominant property at this scale. On the other hand, other models based on hyperbolic/kinetic equations for the evolution of the density of individuals can be assigned, characterized by a finite speed of propagation [6–10].

1.1. The Geometry of the Microfluidic Chip and of the Related Computational Domain

The microfluidic chip is represented as a network of channels connecting two boxes (the microfluidic chambers); see Figure 1 and a schematic picture of the related computational domain is depicted in Figure 2. Here, we refer to the experiment of two main culture chambers (a tumor and an immune cell compartment) connected via narrow capillary migration microchannels with, respectively, width and length of 12 µm and 500 µm. Moreover, the channels height is of 10 µm; however, since in the video footage the experiments is recorded at a fixed height, the third spatial dimension in our framework is neglected. The cross-sectional dimensions of culture chambers are 1 mm (width) × 100 µm (height).

A simplified schematization of the bounded surface where the experiment is performed is reported in Figure 2. We have two microfluidic chambers of the same size, one on the left and the other on the right, defined, respectively, as $\Omega_l = [0, L_x] \times [0, L_y]$ and $\Omega_r := [L_x + L, 2L_x + L] \times [0, L_y]$. They are connected by microchannels, each of them schematized as rectangles $R = [0, L] \times [a, b]$. In order to ease the reading, we point out that in the sequel we approximate the rectangular microchannels as 1D intervals $C = [0, L]$ with zero thickness for the following reasons:

- **modelling reason** the width of microchannels (12 µm) is comparable to the size of cells (for instance, immune cells measure about 8–10 µm of diameter);
- **computational reason** to reduce the running time of the simulation algorithm, since otherwise we should consider a 2D meshgrid for each microchannel.

Then, the link between the box on the left and the corridor is schematized as a junction (node 1L) and analogously the link between the corridor and the box on the right as node 2L. The two junctions are not really a single point, therefore they are parametrized as an interval for node 1L and node 2L—namely, $[a, b]$ of length $\sigma := b - a$.

We remark that for the sake of simplicity, the numerical treatment is developed for a simpler geometry composed by 2D chambers connected through a single 1D channel. The extension to multiple 1D channels is done in Section 3.2.2.

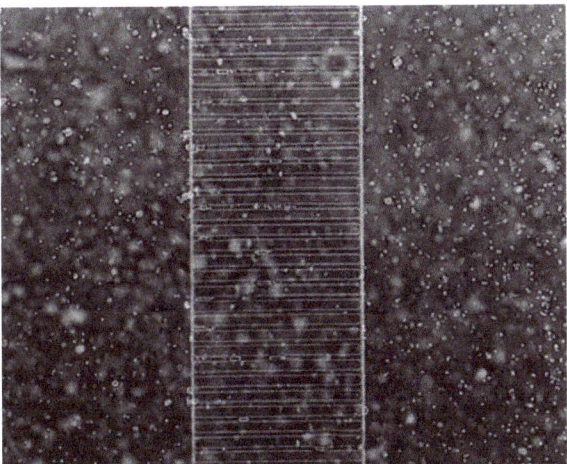

Figure 1. Microfluidic chip environment: two chambers connected by multiple channels. Credits by Vacchelli et al. [1] edited by AAAS.

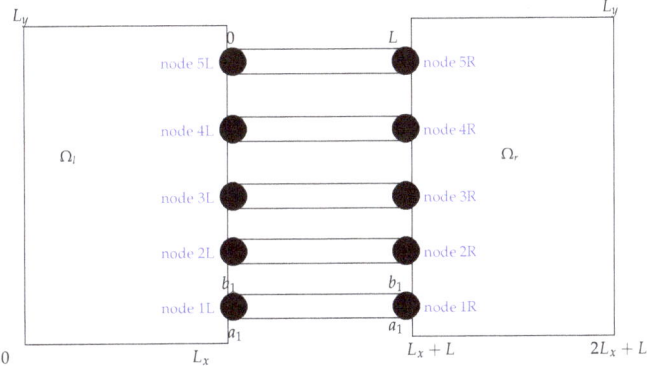

Figure 2. Simplified schematization of the chip geometry depicted in Figure 1.

1.2. Original Contribution of the Present Paper

From the mathematical and numerical viewpoint, here we deal with a challenging issue arising in the chemotaxis modelling of cell interaction. The problem involves doubly parabolic models in 2D domains (microfluifidic chambers) that are connected with 1D domains represented by channels, where either a doubly parabolic or a hyperbolic-parabolic model can be assigned.

The classical doubly parabolic Keller–Segel (KS) model [4] of chemotaxis reads as:

$$\begin{cases} u_t = div(\nu \nabla u - \chi(\phi) u \nabla \phi) \\ \phi_t = D\Delta\phi + au - b\phi, \end{cases} \quad (1)$$

with u the density of individuals in the considered medium, ν the diffusion rate of the organism according to Fick's Law, and ϕ the density of the chemoattractant. The positive constant D is the diffusion coefficient of the chemoattractant; the positive coefficients a and b are, respectively, its production and degradation rates; and χ is the chemotactic sensitivity, depending on the density of the considered quantities. In the 2D domains given by the microfluidic chambers, we apply a reaction-diffusion chemotaxis KS-like model inspired by (1) and described in Section 2.2.1.

In the 1D microfluidic channels, we use the one-dimensional version of the KS-like model used in the chambers, but we also study the behavior of individuals when a hyperbolic-parabolic model, characterized by finite speed of propagation, is assigned. Such a hyperbolic-parabolic model, described in Section 2.2.1, is inspired by the Greeberg–Alt (GA) model, arising as a simple model for chemotaxis on a line:

$$\begin{cases} \partial_t u + \partial_x v = 0, \\ \partial_t v + \lambda^2 \partial_x u = -v + \chi(\phi) u \partial_x \phi, \\ \partial_t \phi = D \partial_{xx} \phi + au - b\phi. \end{cases} \quad (2)$$

Note that here v is the averaged flux. Let us underline that the flux v in model (2) corresponds to $v = -\lambda^2 \nabla u + \chi(\phi) u \nabla \phi$ for the KS system.

We remark that here we do not consider the GA model on the 2D domain, since in this case the derivation of the monotonicity condition, easily computed in 1D, is not straightforward, due to the oscillations brought by 2D wave equation. One possibility to overcome this problem should be to consider the macroscopic hyperbolic model proposed by Preziosi in [9,11]. Alternatively, a kinetic 2D model of chemotaxis and its numerical approximation is studied in [12].

This system was analytically studied on the whole line and on bounded intervals in [13], while an effective numerical approximation, the Asymptotic High Order (AHO) scheme, was introduced in [14]—see also [15,16]—and extended on networks with general boundary conditions in [17,18].

Here, in the numerical treatment for the computation of numerical solutions one has to take care of what happens at the interface when switching from 2D-doubly parabolic models to 1D-doubly parabolic or 1D-hyperbolic-parabolic ones and vice versa.

Since we aim at reproducing the numerical solutions of such models, we need to deal with a multi-domain problem given by the passage from a 2D domain represented by the chambers of the chip to 1D domains given by the channels. For this reason we need to develop ad hoc transmission conditions to ensure mass conservation at the 2D-1D interfaces. From the numerical viewpoint, here we consider numerical boundary conditions including in the stencil a ghost cell value taken from the neighbouring domain, as we will show in the numerical Section 3. The approximation of doubly parabolic chemotaxis models for the 1D-KS model (1) on networks was already considered in [19]. However, in that case the transmission conditions were between 1D–1D interfaces and on each arc of the network the same fully parabolic model was considered. We also underline that in such work, transmission conditions required to impose the continuity of the density of both cells u and chemoattractant ϕ, while we only impose the continuity of the fluxes, which seems to be more realistic when dealing with flux of individuals or molecules.

For the numerical approximation of the GA system (2), we refer to our previous papers [14] for a single line. In particular, the numerical treatment of the hyperbolic part of the system is based on the finite difference AHO scheme with the development of mass-preserving numerical scheme at outer boundaries, while the parabolic part is approximated by finite difference and Crank–Nicolson scheme.

In [17,18] the GA system was solved on networks, thus making it necessary to develop mass-preserving transmission conditions at inner nodes (interfaces) and suitable boundary conditions at outer nodes. However, rgw transmission conditions considered there involved the mass exchange only between 1D–1D interfaces; moreover, on each arc of the network, the same model was considered. Furthermore, the second-order numerical approximation of the boundary conditions developed in such papers did not ensure the posivity preserving property in the case of oscillating functions.

The numerical approximation of permeability Kedem–Katchalsky [20] conditions describing the conservation of the flux through a node was already considered in [21], but we underline that in the mentioned paper the study was done for the approximation with finite elements methods of linear problems. For reaction-diffusion problems the approximation of permeability conditions was studied in [22] for finite difference schemes and in [23] for discontinuous Galerkin methods. The numerical treatment of permeability conditions for chemotaxis problems was presented for the first time in [24] for the 1D parabolic–parabolic interface, and a finite difference approximation was developed without taking into consideration the mass preservation nor the positivity-preservation properties at the interfaces.

Therefore, to the best of our knowledge the present paper is the first numerical work where this new technique of switching the size of the domains and type of equations (parabolic vs. hyperbolic approach) is introduced, in order to develop mass-preserving and positivity preserving schemes. We point out that in the present paper the approximation method—based on finite difference schemes—involves the derivation of suitable zero-flux boundary conditions. The motivation for our framework based on finite difference methods seems to be a natural choice when dealing with hyperbolic models of transmission.

1.3. Main Contents and Plan of the Paper

In the present paper, a positivity-preserving and mass-preserving numerical discretization of Neumann boundary conditions at the corners and at the bottom and top boundaries of the 2D domain for a 2D-doubly parabolic reaction-diffusion problem are presented. Moreover, a positivity-preserving and mass-preserving numerical scheme at the interfaces of the network connecting the 2D chambers with the 1D channels (where the 1D-doubly parabolic or 1D-hyperbolic-parabolic problem can be assigned) is developed. To summarize the main contents of the present work, the mathematical issues faced in this study are categorized into two aspects:

- the study of the behavior of two different modelling of the dynamics in the channels: the parabolic model describing the dynamics inside the chambers was coupled both with KS-like and GA-like models;
- the numerical approximation of equations defined in a heterogeneous domain, characterized by the switch from 2D domains, represented by microfluidic left and right chambers, to 1D domains, given by the channels connecting them.

Then, the numerical questions arising in the mentioned issues and here addressed are:

- the study of positivity and mass-preserving external boundary conditions for 2D-doubly parabolic model (3);
- the introduction of mass-preserving and positivity-preserving permeability conditions at the interfaces between 2D and 1D parabolic models—see Section 2.3.2;
- the introduction of mass-preserving and positivity-preserving permeability conditions at the interfaces between the 2D-fully parabolic model and 1D-hyperbolic-parabolic model—see Section 2.3.3.

The plan of the paper is as follows. In Section 2 we describe the biological framework that inspired our study and we introduce the mathematical formulation of biologically inspired models and we introduce the adopted model. Section 3 is devoted to the numerical techniques used to approximate the problem and in Section 4 some numerical tests showing the qualitative behavior of cells in the designed environment are presented. Finally, in Section 5 a discussion on the results and the future developments of our work is presented.

2. Materials and Methods

The present Section is devoted to:

1. the description of the biological framework and the laboratory experiment that inspired our work—see Section 2.1;
2. the mathematical methods—see Section 2.2—bringing us to the development of a simulation algorithm designed for qualitatively reproducing the experimental observations.

2.1. Biological Framework

The control of immune cells migration and interaction with tumor cells living inside the chambers of the microfluidic chip, represent a new and attractive approach for the clinical management of tumor diseases. Furthermore, in the chip environment also drug testing can be exploited. Then, the quantitative assessment of immune cell migration ability to recognize and attack the tumor cells for each patient could provide a new potential parameter predictive of patient outcomes in the future.

Migrating cells respond to complex chemical stimuli (as a mixture of growth factors, cytokines, and chemokines), representing a source of chemoattractants. These chemoattractants, through the interaction with their receptors, allow cells to acquire a polarized morphology and to perform the action of immunosurveillance.

The development of lab-on-chip technologies has made it possible to realize a reproducible tailoring of the cellular microenvironment, thus allowing the continuous monitoring of experiments and the accurate control of experimental parameters. Recently, the development of microengineering has provided the possibility to realize culturing of multiple cell types and made it possible to observe cell-cell interactions and to transpose in vivo studies to a second generation of in vitro smart environments. The main advantages of this new technological tool are a close control over local experimental conditions and lower costs with respect to the use of animals in laboratory experiments for efficacy and toxicity testing. Some results obtained with on-chip experiments are presented in [1,2,25–27].

Regarding the structure of microfluidic devices, they are designed to allow chemical and physical contacts between tumor cells and non-adherent immune cells (i.e., murine splenocytes or human peripheral blood mononuclear cells). The microfluidic co-culture platforms are fabricated in polydimethylsiloxane (PDMS, Silgard 184), a biocompatible optically transparent silicone elastomer.

2.1.1. Setting of the Laboratory Experiments

Here, we shortly describe the laboratory experiments inspiring our work (see [1]) and carried out in the microfluidic chip environment; see Figure 1. Two populations, immune cells of wild type and cancer cells, are initially put into two separate chambers and can have physical and chemical contact through the microchannels. In more detail, rgw cells are loaded into the reservoirs as follows: the left chambers are filled with about 2×10^6 human peripheral blood mononuclear cells and the right chambers with about 5×10^4 breast cancer cells, pre-treated or not with doxorubicin hydrochloride, all immersed in a suitable culture medium. Time-lapse recordings are performed over a period of 72 h (1 microphotograph every 2 min) by means of a microscope placed directly inside the CO_2 incubator for the duration of the recording.

We consider two different scenarios:

- first scenario (treated case): before enabling cells to migrate, tumor cells are previously treated with a chemoterapy drug. Afterwards, we observe immune cells migrating towards the left chamber where the tumor remains confined, but expresses the chemical stimuli attracting immune cells. Mainly, the dynamics observed in this case is the migration of immune cells from the right to the left in order to attack the tumor cells.
- second scenario (untreated case): tumor cells migrate in the right chamber and proliferate. In this case, the tumor cells do not produce chemoattractant, thus immune cells move in the environment without recognizing and attacking tumor cells.

The culture medium in both cases is neutral, thus meaning that no exogenous substances are introduced. Our aim is then to build a simulation algorithm based on the mathematical model, which is able to reproduce the main features of the observed phenomena in the two scenarios.

2.2. Mathematical Framework

For the development of the mathematical modelling explained in the next section, we remark that we neglect the third dimension, since we do not have laboratory measurements of the movement of cells in the vertical direction. For this reason we consider the microfluidic chambers as 2D objects. Nowadays, the mathematical analysis of biological phenomena has become an important tool to explore complex processes and to detect mechanisms that might not be evident to the experimenters. Although a mathematical model cannot replace a real experiment, it may represent a support tool to explain acquired biological data and it may allow to gain a deeper understanding of the interactions between cancer cells and the immune system. More generally, mathematical models can describe a broad variety of biological phenomena, including cell dynamics and cancer [28–32].

The movement of bacteria under the effect of a chemical substance has been widely studied in the last few decades, and numerous mathematical models have been proposed. As shown in [33], chemotaxis is decisive in biological processes. For instance, the formation of cells aggregations (amoebae, bacteria, etc) occurs during the response of the different species to the change of the chemical gradients in the environment. Moreover it is possible to describe this biological phenomenon at different scales: particles models, hybrid (multiscale) models and macroscopic models.

In the present paper, the population density is assumed to be as a whole, thus macroscopic models of partial differential equations are considered. In particular, in order to describe the dynamics of cells in the 2D chambers we use a KS-like model, while in the microchannels we compare the behavior between two different modelization: a 1D KS-like model and a 1D GA-like model. The modelling here applied is described in the next Section 2.2.1.

2.2.1. The Model

Here, we introduce a mathematical model that aims at describing the behavior of two populations of cells coexisting together: tumoral cells T and immune cells (macrophages) M. We underline that the setting here considered can be made more complex with the introduction of a greater number of cell species and with the presence of an exogenous substance in the environment.

The model consists of a reaction-diffusion system with chemotaxis that it is able to describe birth and death processes, interactions with chemoattractants, interactions and competition between different cell species. The microfluidic chip is schematized as a network of channels connecting two boxes (the microfluidic chambers), then, following the ideas in [17], ad hoc transmission conditions were introduced to ensure mass conservation. The parameters of the model, such as the velocity of different cell populations, the turning rates, and the decay rates, will be calibrated with rgw observed data.

Cancer cells T produce chemical signals, called φ, activating the immune response of M and influencing their behavior. Moreover, we take into account the presence of cytokines ω (produced by M), acting as a chemical killer of cancer cells. Mainly referring to the KS model, the model here considered in the 2D chambers reads as:

$$\begin{cases} \frac{\partial}{\partial t}T = D_T \Delta T - \lambda_T(\omega)T, \\ \frac{\partial}{\partial t}M = D_M \Delta M - div(\chi(\varphi)M \nabla \varphi), \\ \frac{\partial}{\partial t}\varphi = D_\varphi \Delta \varphi + \alpha_\varphi T - \beta_\varphi \varphi, \\ \frac{\partial}{\partial t}\omega = D_\omega \Delta \omega + \alpha_\omega M - \beta_\omega \omega. \end{cases} \qquad (3)$$

The system above describes the dynamics of the two cell species and the diffusion of the chemoattractant, and it needs to be complemented with suitable initial conditions and boundary conditions for the cells and the chemoattractant concentrations, as will be specified in the following paragraphs.

In particular, bearing in mind the first scenario (treated tumor), the system above describes the following situation: tumor cells T produce a chemical substance φ attracting immune cells M, that start migrating towards them. In this case, the tumor does not proliferate, since it is treated by a chemotherapy medicine, thus we do not include this feature in the model. Immune cells do not seem to proliferate during the experiment, thus we neglect this feature in the model. Note that, although here we are neglecting the proliferation of tumor, a tumor growth is observed experimentally in the second scenario (untreated tumor). This feature can be easily added to the model by putting a linear or logistic source term in the equation for the tumor.

Immune cells also produce a chemical substance ω which should be responsible for tumor killing. Therefore, in the first equation of the system (3), besides the diffusion term we can find $-\lambda_T(\omega)T$, representing the tumor suppression operated by immune cells. We underline that at the moment we have no information from the biologists about the real killing rate in the microchip environment induced by ω, but we decided to introduce it in order to include this phenomenon in the model qualitatively.

In the second equation, in addition to the diffusion term we have the chemotactic term $f = \chi(\varphi)\nabla\varphi$ due to the presence of the chemical substance φ produced by the tumor.

In order to define the action of the cytotoxin ω produced by immune cells (which determines the death of cancer cells), we introduce the function $\lambda_T(\omega)$:

$$\lambda_T(\omega) = \frac{S\omega}{\gamma + \omega}, \tag{4}$$

where S is the maximum secretion rate of the cytotoxin by the immune cells and γ the equivalent Michaelis constant associated with the production, as described in [33]. A lot of effort has been devoted to finding a biologically accurate expression for the chemotaxis function $\chi(\varphi)$ representing the chemotactic sensitivity of immune cells; here, we refer the form suggested in [34] by Lapidis and Schiller:

$$\chi(\varphi) = \frac{k_1}{(k_2 + \varphi)^2}, \tag{5}$$

where k_1 represents the cellular drift velocity, while k_2 is the receptor dissociation constant, which says how many molecules are necessary to bind the receptors. Note that we mainly refer to [33] for the values of the parameters k_1, k_2, and all the parameters of the model are reported in Table 1.

Now, in order to describe the dynamics of cells in the microchannels connecting the two boxes, we introduce the following 1D models for the dynamics, with the label c indicating the channels. Then, we consider two possible models in the channels: a diffusive one:

$$\begin{cases} \frac{\partial}{\partial t}T_c = D_T \partial_{xx} T_c - \lambda_T(\omega) T_c, \\ \frac{\partial}{\partial t}M_c = D_M \partial_{xx} M_c - \partial_x(M_c f_c), \\ \frac{\partial}{\partial t}\varphi_c = D_\varphi \partial_{xx} \varphi_c + \alpha_\varphi T_c - \beta_\varphi \varphi_c, \\ \frac{\partial}{\partial t}\omega_c = D_\omega \partial_{xx} \omega_c + \alpha_\omega M_c - \beta_\omega \omega_c, \end{cases} \tag{6}$$

and a hyperbolic one:

$$\begin{cases} \partial_t T_c + \partial_x v_c^T = -\lambda_T(w) T_c, \\ \partial_t v_c^T + D_T \partial_x T_c = -v_c^T, \\ \partial_t w_c = D_w \partial_{xx} w_c + \alpha_w T_c - \beta_c w_c, \\ \partial_x M_c + \partial_t v_c^M = 0, \\ \partial_t v_c^M + D_M \partial_x M_c = M_c f_c - v_c^M, \\ \partial_t \varphi_c = D_\varphi \partial_{xx} \varphi_c + \alpha_\varphi T_c - \beta_\varphi \varphi_c, \end{cases} \quad (7)$$

with T_c and M_c, respectively, as the density of tumor and immune cells; $f_c = \chi(\varphi_c)\partial_x \varphi_c$ and v_c^T and v_c^M, respectively, as the average flux of tumor cells T_c and immune cells M_c in the channels. Note that the 1D-doubly parabolic model (6) is the one-dimensional version of the system (3), while (7) is the 1D-hyperbolic-parabolic model inspired by GA model [5]. We remark that, for the hyperbolic-parabolic system (7), we also need to assign initial and boundary conditions for the flux v.

2.2.2. The Simplified Model

To present the numerical scheme, we write a simpler model with respect to (3) but share the main features of it:

$$\begin{cases} \partial_t u = D_u \Delta u - \mathrm{div} \mathbf{F} + g(x, y, t, u), \\ \partial_t \phi = D_\phi \Delta \phi + au - b\phi, \end{cases} \quad (8)$$

with u as the density of individuals, ϕ as the density of chemoattractant, and with $\mathbf{F} = \mathbf{u}\, f$. From now on, the two components of the drift term $\mathbf{f} = \chi(\phi)\nabla \phi$ will be indicated as:

$$\mathbf{f}(\mathbf{x}, \mathbf{y}, \mathbf{t}) := \begin{pmatrix} f^x(x,y,t) \\ f^y(x,y,t) \end{pmatrix}.$$

For the mono-dimensional channel, in order to make the explanation of the numerical approximation easier, we consider simpler models sharing the same characteristics of the models in (6) and (7), which read as:

$$\begin{cases} \partial_t u_c = D_{u_c} \partial_{xx} u - \partial_x F_c + g(x, t, u), \\ \partial_t \phi_c = D_{\phi_c} \partial_{xx} \phi_c + a_c u_c - b_c \phi_c \end{cases} \quad (9)$$

or

$$\begin{cases} \partial_t u_c + \partial_x v_c = g(x, t, u_c), \\ \partial_t v_c + \lambda_c^2 \partial_x u_c = F_c - v_c, \\ \partial_t \phi_c = D_{\phi_c} \partial_{xx} \phi_c + a_c u_c - b_c \phi_c, \end{cases} \quad (10)$$

with $F_c = u_c f_c$. The systems above have to be complemented with smooth initial conditions for the unknowns u, ϕ and also v for system (10); initial data will be specified in Section 3.3. On the boundary, we consider for all the quantities homogeneous Neumann conditions, so we assume no-flux boundary conditions.

Monotonicity conditions. We also mention at this point that model (10) requires an analytical monotonicity criteria; see [35]. For a linear convection term Au_c and a linear source term $g = Bu_c$, the sub-characteristic criteria

$$\left| \frac{A}{\lambda_c} \right| - B \leq 1,$$

must be satisfied in order for the quantity u_c to be non-negative. Otherwise, having a negative u_c would lead to unphysical solutions.

With regard to our former model (7), that would mean we have for the immune cell density M the monotonicity condition:

$$\frac{k_1}{(k_2 + \varphi_c)^2}|\partial_x \varphi_c| \leq \sqrt{D_M}. \tag{11}$$

This needs to be verified in the computational domain in order to ensure non-negative solutions.

Remark 1. *We remark that the no-flux conditions boundary conditions used in our simulations are needed to have the mass-conservation of all the quantities. However, they are not realistic, since in the laboratory experiment there is an inflow of cells from the outer boundaries. Then, we aim at extending the no-flux boundary conditions to more general ones.*

In the following section, in order to discuss mass-conservation we restrict our study on a 2D closed rectangular domain named Ω_{nc} (i.e., box without the channels) and on single lines not connected at the outer boundaries, as indicated by C_{nc}.

2.3. Outer Boundary and Interface Conditions for the Models with Null Source Term G

From now on, we consider a further simplified version of the models presented above putting the source term g equal to zero to discuss mass conservation.

2.3.1. Boundary Conditions for the 2d Doubly Parabolic Model (8) with $G = 0$.

Here, we consider the mass conservation of cell density u for zero-flux conditions at the outer boundaries of a rectangular closed domain Ω_{nc} for the 2D model (8) with a null source term.

Indeed, since our model describes the migration of cells by both diffusion and chemoattractant effects, physically speaking the mass of cells must be preserved in the absence of the creation and destruction of cells.

For this reason, we assume a no-flux condition for the density u and the chemoattractant ϕ. Since we define the source term **f** as a product function of $\nabla \phi$, we get the no-flux conditions:

$$\mathbf{F}(\mathbf{x}, \mathbf{y}, t)n|_{\partial \Omega_{nc}} = 0, \quad \nabla u(x, y, t)n|_{\delta \Omega_{nc}} = 0, \quad (x, y) \in \partial \Omega_{nc}, \tag{12}$$

These guarantee mass conservation.

2.3.2. Interface between 2D-1D Models in (8) and (9)

We remark that, for simplicity, the following description focuses on the left part of the domain—i.e., Ω_l and its connection with a single microchannel represented by the interval $[0, L]$. However, the numerical treatment holds analogously on the right side of the domain and also to multiple channels.

In the 2D left box Ω_l, the position of node 1L is at $x = L_y, y \in [a, b]$ and for the 1D domain, represented by a given number of channels C, node 1L is placed at $x = 0$ (left endpoint of the channel); see Figure 2.

In order to ensure the conservation of the total mass when the mass-exchange occurs, we introduce suitable transmission conditions at the interface between 2D and 1D domains. Then, we consider the simplified models (9) or (10) for 1D channels, with $g = 0$.

In particular, we have to prescribe the flux conservation at the 2D-1D interfaces, since we cannot lose or gain any cells during the passage through a node.

The conservation condition for u reads as:

$$\frac{d}{dt}\int_{\Omega_l} u(x, y, t)d\Omega_l + \frac{d}{dt}\int_0^L u_c(x, t)dx = 0,$$

and it rewrites as:

$$0 = \oint_{\partial \Omega_l} (D_u \nabla u(x,y,t) - F(x,y,t)) n dS$$
$$+ (D_{u_c} \partial_x u_c - F_c(x,t))|_L^0$$

by using the divergence theorem in the first integral. With our analytical boundary conditions, the integral vanishes, except at the boundary where the node is positioned.

We remark that attention has to be paid to n being the outer normal of the domain. Thanks to the boundary conditions (12), we get the condition:

$$\int_a^b (D_u \partial_x u(L_x, y, t) - u(L_x, y, t) f^x(L_x, y, t)) dy = D_{u_c} \partial_x u_c(0, t) - F_c(0, t). \tag{13}$$

Note that in this case, the evaluation of the integrand at the right endpoint L is discarded, since we are only considering the junction connecting Ω_l with the left endpoint $x = 0$ of the microchannel (see Figure 2).

Now, we impose Kedem–Katchalsky (KK) [20] conditions describing the conservation of the flux through a node (see also [21] for the numerical treatment of these conditions). In particular, at the interface between the left chamber and the channels, we have (on the left of node 1L in Figure 2):

$$D_u \partial_x u(L_x, y, t) - u(L_x, y, t) f^x(L_x, y, t) = K(u_c(0, t) - u(L_x, y, t)) \quad \text{for } y \in [a, b] \tag{14}$$

On the right of node 1L, we have:

$$D_{u_c} \partial_x u_c(0, t) - F_c(0, t) = K u_c(0, t)(b - a) - \int_a^b u(L_x, y, t) dy. \tag{15}$$

Thanks to conditions (14) and (15), we are guaranteed to have the flux conservation (13), and we will use such conditions to obtain numerical boundary conditions for the boundary values at the nodes on both sides, as shown in Section 3 and in Section 3.1.2.

2.3.3. Interface between 2D-1D Models in (8)–(10)

In this section, we describe the combination of 2D parabolic-1D hyperbolic model in order to describe the dynamics with a hyperbolic model in the channels. Further care has to be given in order to keep some important properties which ensure the consistency and non-negativity of numerical solutions when connecting both models.

Now, the transmission conditions for the switch from Ω_l to $C = [0, L]$ are derived in this case. For the mass conservation of u, we impose the condition:

$$0 = \frac{d}{dt} \int_{\Omega_l} u(x, y, t) d\Omega_l + \frac{d}{dt} \int_0^L u_c(x, t) dx$$
$$= \int_{\Omega_l} (D_u \Delta u(x, y, t) - \text{div} F(x, y, t)) d\Omega_l + \int_0^L -\partial_x v(x, t) dx$$
$$\Longrightarrow \oint_{\partial \Omega_l} (D_u \nabla u(x, y, t) - F(x, y, t) n) dS + v(0, t) = 0.$$

Note that in the above formula, we have $v(L, t) = 0$ because we are looking at left interface (node 1L). Then, we finally get:

$$\int_{a_1}^{b_1} (D_u \partial_x u(L_x, y, t) - u(L_x, y, t) f^x(L_x, y, t)) dy = -v(0, t). \tag{16}$$

Now, we impose the KK-condition at the interface:

$$D_u \partial_x u(L_x, y, t) - u(L_x, y, t) f^x(L_x, y, t) = K(u_c(0, t) - u(L_x, y, t)) \quad \text{for } y \in [a, b]$$

Then, (16) reads as:

$$v(0,t) = -K(b-a)u_c(0,t) + K\int_a^b u(L_x,y,t)dy. \tag{17}$$

3. Numerical Approximation

In this section, we describe the numerical approximation of the adopted models: 2D-doubly parabolic, 1D-doubly parabolicm and 1D-hyperbolic-parabolic. We define equispaced $x_i := i\triangle x$, $y_j := j\triangle y$, and $t_n := n\triangle t$ with $\triangle x, \triangle y, \triangle t > 0$, and $i = 0,\ldots,N_x + 1$, $j = 0,\ldots,N_y + 1$; for channel $[0,L]$, we discretize it as $x_i = i\triangle x$, with $i = 0,\ldots,N$. For a more structured presentation, we introduce the operators:

$$\delta_x^2 u_{i,j}^n := u_{i+1,j}^n - 2u_{i,j}^n + u_{i-1,j}^n, \quad \delta_y^2 u_{i,j}^n := u_{i,j+1}^n - 2u_{i,j}^n + u_{i,j-1}^n,$$
$$\delta_x^0 u_{i,j}^n := u_{i+1,j}^n - u_{i-1,j}^n, \quad \delta_y^0 u_{i,j}^n := u_{i,j+1}^n - u_{i,j-1}^n,$$
$$\delta_x^1 u_{i,j}^n := u_{i+1,j}^n - u_{i,j}^n, \quad \delta_y^1 u_{i,j}^n := u_{i,j+1}^n - u_{i,j}^n.$$

Remark 2. *Note that special attention has to be paid also to the stiffness induced by the source term $g(x,y,t,u)$. To overcome this issue, implicit methods can be used, such as the Crank–Nicolson method, which is associated with a Δt that is small enough.*

Mass-preserving and (numerically)-positivity-preserving approximations will be developed in the present section. In the following, we will neglect the label c to make the reading easier and we will make distinctions only when necessary.

3.1. The Parabolic-Parabolic Case

Here, we introduce a numerical scheme for the doubly parabolic systems (8) and (9) for $g = 0$.

For the discretization of equations in a 2D system (8) in the interior points of the domain—i.e., for $i = 1,\ldots,N_x, j = 1,\ldots,N_y$, we define a finite difference discretization both for u and ϕ:

$$u_{i,j}^{n+1} = u_{i,j}^n + D_u\frac{\Delta t}{2}\left[\frac{\delta_x^2(u_{i,j}^n+u_{i,j}^{n+1})}{\Delta x^2} + \frac{\delta_y^2(u_{i,j}^n+u_{i,j}^{n+1})}{\Delta y^2}\right] \\ -\Delta t(\delta_x^0 F_{i,j}^{x,n} + \delta_y^0 F_{i,j}^{y,n}), \tag{18}$$

$$\phi_{i,j}^{n+1} = \phi_{i,j}^n + D_\phi\frac{\Delta t}{2}\left[\frac{\delta_x^2(\phi_{i,j}^n+\phi_{i,j}^{n+1})}{\Delta x^2} + \frac{\delta_y^2(\phi_{i,j}^n+\phi_{i,j}^{n+1})}{\Delta y^2}\right] \\ + \frac{\Delta t}{2}a(u_{i,j}^n + u_{i,j}^{n+1}) - \frac{\Delta t}{2}b(\phi_{i,j}^n + \phi_{i,j}^{n+1}). \tag{19}$$

Note that $F_i^{x,n} = \chi(\phi_{i,j}^n)u_{i,j}^n\delta_x^0\phi_{i,j}^n$ with $\delta_x^0\phi_{i,j}^n$ a centered second order approximation of ϕ_x.

For a 1D system (9), in the interior points of the channel C we apply the Crank–Nicolson scheme, as above:

$$u_i^{c,n+1} = u_i^{c,n} + D_{uc}\frac{\Delta t}{2}\frac{\delta_x^2(u_i^{c,n}+u_i^{c,n+1})}{\Delta x^2} - \Delta t\delta_x^0 F_{c,i}^n, \tag{20}$$

$$\phi_i^{c,n+1} = \phi_i^{c,n} + D_{\phi c}\frac{\Delta t}{2}\frac{\delta_x^2(\phi_i^{c,n}+\phi_i^{c,n+1})}{\Delta x^2} \\ + \frac{\Delta t}{2}a_c(u_i^{c,n} + u_i^{c,n+1}) - \frac{\Delta t}{2}b_c(\phi_i^{c,n} + \phi_i^{c,n+1}). \tag{21}$$

Remark 3. *We remark that here, for simplicity, upwinding terms needed to avoid meshgrid restrictions caused by a large Péclet number—see, for instance [36]—are not included in the schemes (18) and (20). In this case, assuming a small Δx and Δy will be enough to avoid oscillations.*

The upwinding be added in the implemented scheme described in Section 3.3.

Stability condition. By using the Von-Neumann stability analysis on the linearized problem, we derive the following condition for the Crank–Nicolson schemes above, [37]:

$$D_u \frac{\Delta t}{\Delta x^2} \leq 1 \quad \text{for 1D,} \quad D_u \frac{\Delta t}{\Delta x^2} + D_u \frac{\Delta t}{\Delta y^2} \leq 1 \quad \text{for 2D.} \tag{22}$$

In the following, we present the discretization of the boundary and transmission conditions to complete the numerical schemes.

3.1.1. Discretization of the Outer Boundary Conditions for the Doubly Parabolic Problem

Since a qualitative characteristic of the model is the preservation of total mass for zero-flux boundary conditions, the first step is to ensure the mass preservation at each time step in a closed 1D line C_{nc} and, analogously, in a closed 2D chamber, namely Ω_{nc}. To this aim, we have to choose discrete boundary conditions that both are consistent with the analytical boundary conditions and preserve the mass in the numerical method. We remark that we present the computations without source term g, and we will add it in the following to complete the equations.

Boundary conditions for the density of individuals u in 1D model (9).

Here, we consider Neumann boundary conditions $\frac{\partial u}{\partial x}(x,t) = 0$. If we discretized it with a (standard) forward finite difference scheme:

$$\partial_x u(0) = \frac{u_1^{c,n} - u_0^{c,n}}{\Delta x}, \tag{23}$$

the mass will not be preserved over time, as shown in the numerical Example 1. Therefore, in order to have mass preservation, we use the central finite difference approximation with a ghost cell:

$$\partial_x u(0) = \frac{u_1^{c,n} - u_{-1}^{c,n}}{2\Delta x}. \tag{24}$$

At the outer boundaries at the first ($x_0 = 0$) and last ($x_{N+1} = L$) endpoint of the channels, we assign the numerical schemes:

$$u_0^{c,n+1} = u_0^{c,n} + D_{u_c} \frac{\Delta t}{\Delta x^2} \delta_x^1 \left(u_0^{c,n} + u_0^{c,n+1} \right) - \frac{\Delta t}{\Delta x} (F_{c,1}^n + F_{c,0}^n) \tag{25}$$

and

$$u_{N+1}^{c,n+1} = u_{N+1}^{c,n} - D_{u_c} \frac{\Delta t}{\Delta x^2} \delta_x^1 \left(u_N^{c,n} + u_N^{c,n+1} \right) + \frac{\Delta t}{\Delta x} (F_{c,N}^n + F_{c,N+1}^n), \tag{26}$$

with $F_{c,0}^n = 0$ in (25) and $F_{c,N+1}^n = 0$ in (26), since we are imposing zero flux at the boundaries. The boundary conditions above are mass-preserving by construction, as stated in the following proposition.

Proposition 1. *The 1D numerical scheme at the internal points introduced above, namely:*

$$u_i^{c,n+1} = u_i^{c,n} + D_{u_c} \frac{\Delta t}{2\Delta x^2} \delta_x^2 (u_i^{c,n} + u_i^{c,n+1}) - \frac{\Delta t}{2\Delta x} \delta_x^0 F_{c,i}^n, \tag{27}$$

endowed with boundary conditions (25) and (26), is mass-preserving by construction, since it is obtained by imposing $\mathcal{I}^{n+1} - \mathcal{I}^n = 0$.

Proof. The mass conservation over time on the closed domain Ω_{nc} reads as:

$$I(t) = \int_{\Omega_{nc}} u(t,x,y) d\Omega_{nc} = \int_{\Omega_{nc}} u(0,x,y) d\Omega_{nc} = I(0). \tag{28}$$

Now, applying a quadrature rule for the numerical integration:

$$\mathcal{I}^n \approx \int_{\Omega_{nc}} u(t,x,y) d\Omega_{nc}. \tag{29}$$

We need to ensure that $\mathcal{I}^{n+1} = \mathcal{I}^n$.

For the numerical integration, different quadrature formulas can be used; in particular, we applied closed Newton–Cotes formulas to take into account the values at the boundaries. By using the trapezoidal rule with an integration error of $\mathcal{O}(\Delta x^2)$ and imposing the equality $\mathcal{I}^{n+1} - \mathcal{I}^n = 0$ in the 1D case, one obtains:

$$\Delta x \left(\frac{u_0^{c,n+1}}{2} - \frac{u_0^{c,n}}{2} + \sum_{i=1}^{N} \left(u_i^{c,n+1} - u_i^{c,n} \right) + \frac{u_{N+1}^{c,n+1}}{2} - \frac{u_{N+1}^{c,n}}{2} \right) = 0.$$

Using the numerical scheme (20) for $u_i^{c,n+1}$ in (30) for $i = 1, \ldots, N$ we have:

$$\Delta x \left(\frac{u_0^{c,n+1} - u_0^{c,n}}{2} + \frac{D_{uc}\Delta t}{2\Delta x^2} \sum_{i=1}^{N} \delta_x^2 \left(u_i^{c,n} + u_i^{c,n+1} \right) \right.$$
$$\left. - \frac{\Delta t}{2\Delta x} \sum_{i=1}^{N} \delta_x^0 F_{c,i}^n + \frac{u_{N+1}^{c,n+1} - u_{N+1}^{c,n}}{2} \right) = 0.$$

Now, using the definition of $\delta_x^0 F_{c,i}^n$ in the sum, the above formula becomes:

$$u_0^{c,n+1} - u_0^{c,n} - D_{uc}\frac{\Delta t}{\Delta x^2} \left(u_1^{c,n} + u_1^{c,n+1} - (u_0^{c,n} + u_0^{c,n+1}) \right) + \frac{\Delta t}{\Delta x}\left(F_{c,0}^n + F_{c,1}^n \right)$$
$$+ u_{N+1}^{c,n+1} - u_{N+1}^{c,n} - D_{uc}\frac{\Delta t}{\Delta x^2} \left(u_N^{c,n} + u_N^{c,n+1} - (u_{N+1}^{c,n} + u_{N+1}^{c,n+1}) \right)$$
$$- \frac{\Delta t}{\Delta x}\left(F_{c,N}^n + F_{c,N+1}^n \right) = 0.$$

We can now compute the values for both u_0^{n+1} and u_{N+1}^{n+1} so that the term equals zero. By collecting values from nearby stencils together (otherwise we obtain an error of $\mathcal{O}(\Delta x)$, which can be verified by Taylor expansion), at the outer boundaries of 1D domain we obtain the schemes (25) and (26). Note that in the formula above, by imposing homogeneous Neumann boundary conditions we have $f_0^n = 0$ and $f_{N+1}^n = 0$.

We also remark that besides imposing the stability condition (22), Δx has to be small enough in order to ensure the positivity of the scheme for taking into account the possibility of having a negative term brought by f_1^n or f_N^n in (25) and (26), respectively. □

Remark 4. *Although here we are not proving that, the scheme (27) is in practice second order in the space up to the boundaries, since it can be equivalently obtained using the second-order approximation of the first derivative including a ghost cell reported in (24).*

We underline that for $\mathbf{f} \neq 0$, even using formula (24), the approach with the discrete integral equation is necessary for ensuring the mass preservation. Furthermore, by using a different numerical integration scheme, we can achieve different mass-preserving boundary conditions of a higher order.

Boundary conditions for the density of individuals u in 2D model (8).

Let us assume there are no-flux conditions at the boundaries. Then, in this case we consider the 2D closed domain Ω_{nc}.

Using the mass-preserving property argument, we compute boundary conditions for the corners and the top and bottom boundaries of Ω_{nc}. By applying them with the numerical method (18) into $\mathcal{I}^{n+1} - \mathcal{I}^n = 0$, we obtain the expression:

$$\frac{\Delta t \Delta x}{4} \left(-4\Delta t \sum_{i=1}^{N_x} \sum_{j=1}^{N_y} \left(\delta_x^0 F_{i,j}^x + \delta_y^0 F_{i,j}^y \right) \right) = 0,$$

since the terms in u cancel. By plugging in it the expression of the central in rgw space second-order finite difference scheme $\delta^0_x f^{x,n}_j$ for $\text{div}(\mathbf{f})$, we obtain:

$$\frac{1}{\Delta y} \sum_{i=1}^{N_x} \left(F^{y,n}_{i,N_y+1} + F^{y,n}_{i,N_y} - F^{y,n}_{i,1} - F^{y,n}_{i,0} \right)$$

$$+ \frac{1}{\Delta x} \sum_{j=1}^{N_y} \left(F^{x,n}_{N_x+1,j} + F^{x,n}_{N_x,j} - F^{x,n}_{1,j} - F^{x,n}_{0,j} \right) = 0.$$

Now, we can distribute the remaining values to the boundaries in the same way as above for the 1D-parabolic case. Therefore, we obtain the following mass-preserving boundary conditions for the corners:

$$\begin{cases} u^{n+1}_{0,0} = u^n_{0,0} + D_u \frac{\Delta t}{2\Delta x^2} \delta^1_x \left(u^n_{0,0} + u^{n+1}_{0,0} \right) + D_u \frac{\Delta t}{2\Delta y^2} \delta^1_y \left(u^n_{0,0} + u^{n+1}_{0,0} \right), \\[4pt]
u^{n+1}_{N_x+1,0} = u^n_{N_x+1,0} - D_u \frac{\Delta t}{2\Delta x^2} \delta^1_x \left(u^n_{N_x,0} + u^{n+1}_{N_x,0} \right) \\
\qquad\quad + D_u \frac{\Delta t}{2\Delta y^2} \delta^1_y \left(u^n_{N_x+1,0} + u^{n+1}_{N_x+1,0} \right), \\[4pt]
u^{n+1}_{0,N_y+1} = u^n_{0,N_y+1} + D_u \frac{\Delta t}{2\Delta x^2} \delta^1_x \left(u^n_{0,N_y+1} + u^{n+1}_{0,N_y+1} \right) \\
\qquad\quad - D_u \frac{\Delta t}{2\Delta y^2} \delta^1_y \left(u^n_{0,N_y} + u^{n+1}_{0,N_y} \right), \\[4pt]
u^{n+1}_{N_x+1,N_y+1} = u^n_{N_x+1,N_y+1} - D_u \frac{\Delta t}{2\Delta x^2} \delta^1_x \left(u^n_{N_x,N_y+1} + u^{n+1}_{N_x,N_y+1} \right) \\
\qquad\quad - D_u \frac{\Delta t}{2\Delta y^2} \delta^1_y \left(u^n_{N_x+1,N_y} + u^{n+1}_{N_x+1,N_y} \right). \end{cases} \qquad (30)$$

For the edges of the box Ω_{nc}, $i = 1, \ldots, N_x$ and $j = 1, \ldots, N_y$, we have:

$$\begin{cases} u^{n+1}_{i,0} = u^n_{i,0} + D_u \frac{\Delta t}{2\Delta x^2} \delta^2_x \left(u^n_{i,0} + u^{n+1}_{i,0} \right) + D_u \frac{\Delta t}{\Delta y^2} \delta^1_y \left(u^n_{i,0} + u^{n+1}_{i,0} \right) \\
\qquad\quad - \frac{\Delta t}{\Delta y} F^{y,n}_{i,1}, \\[4pt]
u^{n+1}_{i,N_y+1} = u^n_{i,N_y+1} + D_u \frac{\Delta t}{2\Delta x^2} \delta^2_x \left(u^n_{i,N_y+1} + u^{n+1}_{i,N_y+1} \right) \\
\qquad\quad - D_u \frac{\Delta t}{\Delta y^2} \delta^1_y \left(u^n_{i,N_y} + u^{n+1}_{i,N_y} \right) + \frac{\Delta t}{\Delta y} F^{y,n}_{i,N_y}, \\[4pt]
u^{n+1}_{0,j} = u^n_{0,j} + D_u \frac{\Delta t}{\Delta x^2} \delta^1_x \left(u^n_{0,j} + u^{n+1}_{0,j} \right) + D_u \frac{\Delta t}{2\Delta y^2} \delta^2_y \left(u^n_{0,j} + u^{n+1}_{0,j} \right) \\
\qquad\quad - \frac{\Delta t}{\Delta x} F^{x,n}_{1,j}, \\[4pt]
u^{n+1}_{N_x+1,j} = u^n_{N_x+1,j} - D_u \frac{\Delta t}{\Delta x^2} \delta^1_x \left(u^n_{N_x,j} + u^{n+1}_{N_x,j} \right) \\
\qquad\quad + D_u \frac{\Delta t}{2\Delta y^2} \delta^2_y \left(u^n_{N_x+1,j} + u^{n+1}_{N_x+1,j} \right) + \frac{\Delta t}{\Delta x} F^{y,n}_{i,N_y}. \end{cases} \qquad (31)$$

We underline that in the above formulas the terms $F^{x,n}$ and $F^{y,n}$ on the edges of the domain cancel because of the homogeneous Neumann boundary condition on ϕ_x and ϕ_y.

Boundary conditions for the density of chemoattractant ϕ in 1D model (9).

For the computation of the conditions at the outer boundaries for the chemoattractant ϕ_c in the 1D-doubly parabolic model we proceed as above, but neglect the source term $a_c u_c - b_c \phi_c$ to obtain boundary conditions that are mass-preserving.

Proceeding as above, we achieve the following mass-preserving boundary conditions for the chemoattractant:

$$\begin{cases} \phi^{c,n+1}_0 = \phi^{c,n}_0 + D_{\phi_c} \frac{\Delta t}{\Delta x^2} \left(\delta^1_x \phi^{c,n}_0 + \delta^1_x \phi^{c,n+1}_0 \right) \\
\phi^{c,n+1}_{N+1} = \phi^{c,n}_{N+1} - D_{\phi_c} \frac{\Delta t}{\Delta x^2} \left(\delta^1_x \phi^{c,n}_N + \delta^1_x \phi^{c,n+1}_N \right) \end{cases} \qquad (32)$$

This, under the CFL condition (22), is second-order accurate and positivity-preserving up to the boundaries. The parabolic equation in the interior points is solved using an implicit-explicit method:

$$\phi_i^{c,n+1} = \phi_i^{c,n} + D_{\phi_c}\frac{\Delta t}{2\Delta x^2}\delta_x^2\left(\phi_i^{c,n} + \phi_i^{c,n+1}\right). \tag{33}$$

Boundary conditions for the density of chemoattractant ϕ in 2D model (8).

Reasoning as above, by applying the numerical method (19) we obtain the following boundary condition for the chemoattractant at the corners:

$$\begin{cases} \phi_{0,0}^{n+1} = \phi_{0,0}^n + D_\phi\frac{\Delta t}{\Delta x^2}\delta_x^1(\phi_{0,0}^n + \phi_{0,0}^{n+1}) + 2D_\phi\frac{\Delta t}{\Delta y^2}\delta_y^1(\phi_{0,0}^n + \phi_{0,0}^{n+1}) \\[4pt] \phi_{N_x+1,0}^{n+1} = \phi_{N_x+1,0}^n - D_\phi\frac{\Delta t}{\Delta x^2}\delta_x^1(\phi_{N_x+1,0}^n + \phi_{N_x+1,0}^{n+1}) \\ \qquad\qquad + D_\phi\frac{\Delta t}{\Delta y^2}\delta_y^1(\phi_{N_x+1,0}^n + \phi_{N_x+1,0}^n) \\[4pt] \phi_{N_x+1,N_y+1}^{n+1} = \phi_{N_x+1,N_y+1}^n - D_\phi\frac{\Delta t}{\Delta x^2}\delta_x^1(\phi_{N_x,N_y+1}^n + \phi_{N_x,N_y+1}^{n+1}) \\ \qquad\qquad - D_\phi\frac{\Delta t}{\Delta y^2}\delta_x^1(\phi_{N_x+1,N_y+1}^n + \phi_{N_x+1,N_y}^n) \\[4pt] \phi_{0,N_y+1}^{n+1} = \phi_{0,0}^n + D_\phi\frac{\Delta t}{\Delta x^2}\delta_x^1(\phi_{1,N_y+1}^n + \phi_{1,N_y+1}^{n+1}) \\ \qquad\qquad - D_\phi\frac{\Delta t}{\Delta y^2}\delta_y^1(\phi_{0,N_y}^n + \phi_{0,N_y}^{n+1}) \end{cases} \tag{34}$$

and

$$\begin{cases} \phi_{i,0}^{n+1} = \phi_{i,0}^n + D_\phi\frac{\Delta t}{\Delta x^2}\delta_x^2(\phi_{i,0}^n + \phi_{i,0}^{n+1}) + 2D_\phi\frac{\Delta t}{\Delta y^2}\delta_y^1(\phi_{i,0}^n + \phi_{i,0}^{n+1}) \\[4pt] \phi_{i,N_y+1}^{n+1} = \phi_{i,N_y+1}^n + D_\phi\frac{\Delta t}{\Delta x^2}\delta_x^2(\phi_{i,N_y+1}^n + \phi_{i,N_y+1}^{n+1}) \\ \qquad\qquad - 2D_\phi\frac{\Delta t}{\Delta y^2}\delta_y^1(\phi_{i,N_y}^n + \phi_{i,N_y}^{n+1}) \\[4pt] \phi_{0,j}^{n+1} = \phi_{0,j}^n + D_\phi\frac{\Delta t}{\Delta y^2}\delta_y^2(\phi_{0,j}^n + \phi_{0,j}^{n+1}) + 2D_\phi\frac{\Delta t}{\Delta x^2}\delta_x^1(\phi_{0,j}^n + \phi_{0,j}^{n+1}) \\[4pt] \phi_{N_x+1,j}^{n+1} = \phi_{N_x+1,j}^n + D_\phi\frac{\Delta t}{\Delta y^2}\delta_y^2(\phi_{N_x+1,j}^n + \phi_{N_x+1,j}^{n+1}) \\ \qquad\qquad - 2D_\phi\frac{\Delta t}{\Delta x^2}\delta_x^1(\phi_{N_x,j}^n + \phi_{N_x,j}^{n+1}) \end{cases} \tag{35}$$

We now have a complete numerical method to solve the system (8) on Ω_{nc} and the 1D version of it (9) on C_{nc}.

Let us now turn to the complete domain depicted in Figure 2 in order to develop mass-preserving transmission conditions at the nodes of the network. We remark that for the sake of clarity, for the development of the numerical transmission conditions here we consider just the junction—indicated as node 1L—connecting the left box Ω_l and a single channel parametrized as an interval $C = [a,b]$.

3.1.2. Discretization of the Transmission Conditions for the 2D-1D Doubly Parabolic Case

The choice of suitable transmission conditions is crucial, since it should reflect the qualitative attributes of the analytical model.

Here, we use ghost values (24) in order to obtain the numerical boundary conditions, since in such a way mass-preserving and positivity-preserving properties are ensured.

By using the central approximation formula for $\operatorname{div} f$ in the condition (14) on the left of node 1L, we have:

$$D_u\partial_x u(L_x,y,t) - F^x(L_x,y,t) = K(u_c(0,t) - u(L_x,y,t)) \text{ for } y \in [a,b].$$

Then we have:

$$D_u \frac{u^n_{N_x+2,j} - u^n_{N_x,j}}{2\Delta x} = K\left(u^{c,n}_0 - u^n_{N_x+1,j}\right) + F^{x,n}_{N_x+1,j}$$

and we get:

$$u^n_{N_x+2,j} = u^n_{N_x,j} + K\frac{2\Delta x}{D}\left(u^{c,n}_0 - u^n_{N_x+1,j}\right) + \frac{2\Delta x}{D_u} F^{x,n}_{N_x+1,j} \qquad (36)$$

for $j = j_a, \ldots, j_b$.

Moreover, using the central approximation for $\partial_x u_c$ in (15), we finally get the formula:

$$u^{c,n}_{-1} = u^{c,n}_1 - K\frac{2\Delta x}{D_{u_c}}(b-a)u^{c,n}_0 + \frac{2\Delta x}{D_{u_c}}\Delta y \sum_{j=j_a}^{j_b} K(u^n_{N_x+1,j} + u^{n+1}_{N_x+1,j}) - \frac{2\Delta x}{D_{u_c}} F^n_{c,0}. \qquad (37)$$

We now use the Ansatz to apply the ghost values into the 1D (27) and 2D (18) numerical schemes without specific chemotactic approximation, and use the discrete integral equation to determine the chemotactic term discretization.

Since we need to conserve the mass in each domain, but also in both connected ones, the expanded discrete integral equation is needed to compute the total mass over both domains.

Plugging the ghost values (36) and (37), respectively, into the numerical schemes (18) and (27), we get the conditions at the interface (node 1L):

$$\begin{aligned}
u^{n+1}_{N_x+1,j} = & u^n_{N_x+1,j} - D_u \frac{\Delta t}{\Delta x^2} \delta^1_x \left(u^n_{N_x,j} + u^{n+1}_{N_x,j}\right) + 2K\frac{\Delta t}{\Delta x}\left(u^{c,n}_0 - u^n_{N_x+1,j}\right) \\
& + D_u \frac{\Delta t}{2\Delta y^2} \delta^2_y \left(u^n_{N_x+1,j} + u^{n+1}_{N_x+1,j}\right) \\
& + 2\frac{\Delta t}{\Delta x} F^{x,n}_{N_x+1,j} - \Delta t \delta^0_x F^{x,n}_{N_x+1,j} - \Delta t \delta^0_y F^{y,n}_{N_x+1},
\end{aligned} \qquad (38)$$

and

$$\begin{aligned}
u^{c,n+1}_0 = & u^{c,n}_0 + D_{u_c} \frac{\Delta t}{\Delta x^2} \delta^1_x \left(u^{c,n}_0 + u^{c,n+1}_0\right) - 2K\frac{\Delta t}{\Delta x}(b-a)u^{c,n}_0 \\
& + 2K\frac{\Delta t \Delta y}{\Delta x} \sum_{j=j_a}^{j_b} \left(u^n_{N_x+1,j} + u^{n+1}_{N_x+1,j}\right) \\
& - 2\frac{\Delta t}{\Delta x} F^n_{c,0} - \frac{\Delta t}{2\Delta x} \delta^0_x F^n_{c,0}.
\end{aligned} \qquad (39)$$

In particular, the conservation of the discrete total mass reads as:

$$\mathcal{I}^{n+1}_{1D} + \mathcal{I}^{n+1}_{2D} - \mathcal{I}^n_{1D} - \mathcal{I}^n_{2D} = 0, \qquad (40)$$

Now, applying the conditions (38) and (39) with the other boundary conditions (31) and (25), we get:

$$\begin{aligned}
& \Delta x\Big(- K\frac{\Delta t}{\Delta x}(b-a)u^{c,n}_0 + K\frac{\Delta t \Delta y}{\Delta x} \sum_{j=j_a}^{j_b}\left(u^n_{N_x+1,j} + u^{n+1}_{N_x+1,j}\right) \\
& - \frac{\Delta t}{\Delta x} F^n_{c,0} - \frac{\Delta t}{2}\delta^0_x F^n_{c,0} + \frac{\Delta t}{2\Delta x}\left(F^n_{c,0} + F^n_{c,1}\right)\Big) \\
& + \frac{\Delta x \Delta y}{4}\Big(2\sum_{j=j_a}^{j_b}\Big(2K\frac{\Delta t}{\Delta x}\left(u^{c,n}_0 - u^n_{N_x+1,j}\right) + 2\frac{\Delta t}{\Delta x} F^{x,n}_{N_x+1,j} - \Delta t \delta^0_x F^{x,n}_{N_x+1,j} \\
& - \Delta t \delta^0_y F^{y,n}_{N_x+1,j}\Big) \\
& - \frac{2\Delta t}{\Delta x}\sum_{j=j_{a_1}}^{j_{b_1}}\left(F^{x,n}_{N_x+1,j} + F^{x,n}_{N_x,j}\right) - \frac{2\Delta t}{\Delta y}\sum_{j=j_a}^{j_b}\left(F^{y,n}_{N_x+1,j} + F^{y,n}_{N_x,j}\right)\Big) = 0,
\end{aligned}$$

We can then obtain the following transmission conditions:

$$u_0^{c,n+1} = \underbrace{u_0^{c,n} + D_{u_c}\frac{\Delta t}{\Delta x^2}\delta_x^1\left(u_0^{c,n} + u_0^{c,n+1}\right) - \frac{\Delta t}{\Delta x}(F_{c,1}^n + F_{c,0}^n)}_{\text{same as for BC without transmission condition}}$$
$$-2K\frac{\Delta t}{\Delta x}(b-a)u_0^{c,n} + 2K\frac{\Delta t}{\Delta x}\sum_{j=j_a}^{j_b} u_{N_x+1,j}^n + u_{N_x+1,j}^{n+1}, \quad (41)$$

$$\begin{aligned}
u_{N_x+1,j}^{n+1} &= u_{N_x+1,j}^n - D_u\frac{\Delta t}{\Delta x^2}\delta_x^1\left(u_{N_x,j}^n + u_{N_x,j}^{n+1}\right) \\
&+ D_u\frac{\Delta t}{2\Delta y^2}\delta_y^2\left(u_{N_x+1,j}^n + u_{N_x+1,j}^{n+1}\right) \\
&+ \frac{\Delta t}{\Delta x}\left(F_{N_x+1,j}^{x,n} + F_{N_x,j}^{x,n}\right) - \frac{\Delta t}{2\Delta y}\left(F_{N_x+1,j+1}^{y,n} - F_{N_x+1,j-1}^{y,n}\right) \\
&+ \underbrace{2K\frac{\Delta t}{\Delta x}\left(u_0^{c,n} - u_{N_x+1,j}^n\right)}_{\text{additional term for transmission condition,}}
\end{aligned} \quad (42)$$

for $j = j_a, \ldots, j_b$.

Proceeding analogously as above, this approach leads to mass-preserving and positivity-preserving transmission conditions for the chemoattractant ϕ as well. In particular, we have at the first and last endpoint, respectively:

$$\begin{aligned}
\phi_0^{c,n+1} &= \phi_0^{c,n} + D_{\phi_c}\frac{\Delta t}{\Delta x^2}\left(\delta_x^1\phi_0^{c,n} + \delta_x^1\phi_0^{c,n+1}\right) + \Delta t a_c u_0^{c,n} - \Delta t b_c \phi_0^{c,n} \\
&- 2K\frac{\Delta t}{\Delta x}(b-a)\phi_0^{c,n} + 2K\frac{\Delta t}{\Delta x}\int_a^b \phi(L_x, y, t_n)dy
\end{aligned}$$

and

$$\begin{aligned}
\phi_{N_x+1,j}^{n+1} &= \phi_{N_x+1,j}^n + D_\phi\frac{\Delta t}{\Delta y^2}\delta_y^2(\phi_{N_x+1,j}^n + \phi_{N_x+1,j}^{n+1}) \\
&- 2D_\phi\frac{\Delta t}{\Delta x^2}\delta_x^1(\phi_{N_x,j}^n + \phi_{N_x,j}^{n+1}) \\
&+ a\Delta t u_{N_x+1,j}^n - b\Delta t \phi_{N_x+1,j}^n + 2K\frac{\Delta t}{\Delta x}\left(\phi_0^{c,n} - \phi_{N_x+1,j}^n\right).
\end{aligned} \quad (43)$$

We have finally developed a complete numerical scheme to treat doubly parabolic partial differential equations systems in two domains, 1D and 2D, connected through a node, which ensures by construction the mass conservation as the original PDE. Numerically, the scheme also ensures the positivity preserving property under the monotonicity conditions discussed in Section 3.3.1.

3.2. The Hyperbolic-Parabolic Case

The second-order AHO scheme on a line was introduced in [14] for the 1D hyperbolic system (2). Here, considering the presence of the source term g on the right hand side of the equation for the density of cells, the AHO scheme of second order reads as:

$$\begin{cases}
\begin{aligned}
u_i^{c,n+1} &= u_i^{c,n} + \lambda\frac{\Delta t}{2\Delta x}\delta_x^2 u_i^{c,n} - \left(\frac{\Delta t}{2\Delta x} - \frac{\Delta t}{4\lambda}\right)\delta_x^0 v_i^{c,n} + \frac{\Delta t}{4\lambda}\delta_x^0 \Gamma_{c,i}^n \\
&+ \frac{\Delta t}{4}(g_{i-1}^n + 2g_i^n + g_{i+1}^n),
\end{aligned} \\
\begin{aligned}
v_i^{c,n+1} &= v_i^{c,n} - \lambda^2\frac{\Delta t}{2\Delta x}\delta_x^0 u_i^{c,n} + \left(\frac{\lambda\Delta t}{2\Delta x} - \frac{\Delta t}{4}\right)\delta_x^2 v_i^{c,n} + \frac{\Delta t}{4}\delta_x^2 F_{c,i}^n \\
&+ \lambda\frac{\Delta t}{4}(g_{i-1}^n - g_{i+1}^n),
\end{aligned}
\end{cases} \quad (44)$$

with mass-preserving boundary conditions (including the additional source term g) at the external boundaries.

We remark that for the hyperbolic-parabolic model not only mass must be preserved as the in the fully parabolic model, but also the flux v needs to converge towards the steady state $v = 0$. Since here we have the 1D domain connected at both the endpoints, we do not need to use numerical boundary conditions for the outer boundaries. However, for the details and the description of the AHO numerical scheme at the outer boundaries, see [14]. For this reason we use the so-called AHO (Asymptotic Higher Order) schemes (see [18] for the study of AHO scheme at interfaces including mass-preserving transmission conditions) with source term g, for which the approximation of the stationary solutions is up to the third order of accuracy and converges towards a numerical solution with $v = 0$, while preserving the mass.

3.2.1. Discretization of Transmission Conditions for the 2D-Doubly Parabolic and 1D-Hyperbolic-Parabolic Case

The first equation is the same as for the interface between the 2D-doubly parabolic and 1D-doubly parabolic case. Hence, we derive the same transmission condition reported in (42) for $u_{N_x+1,j}^{n+1}$ for $j = j_a, \ldots, j_b$.

For the flux, the transmission condition (17) gives us

$$v_0^{c,n+1} = -K(b-a)u_0^{c,n+1} + K\Delta y \sum_{j=j_a}^{j_b} \left(u_{N_x+1,j}^n + u_{N_x+1,j}^{n+1} \right), \quad (45)$$

This can be computed explicitly with the numerically computed values of $u_0^{c,n+1}$ and $u_{N_x+1,j}^{n+1}$. Then, imposing the mass conservation:

$$\mathcal{I}_{2D}^{n+1} + \mathcal{I}_{2D}^{n+1} - \mathcal{I}_{1D}^n - \mathcal{I}_{1D}^n = 0$$

we get:

$$\frac{\Delta x}{2}\left[u_0^{c,n+1} - u_0^{c,n} + \lambda \frac{\Delta t}{\Delta x}\left(u_0^{c,n+1} - u_1^{c,n+1} \right) - \left(\frac{\Delta t}{\Delta x} - \frac{\Delta t}{2\lambda} \right)\left(-v_0^{c,n+1} - v_1^{c,n+1} \right) \right]$$
$$+ \frac{\Delta x \Delta t}{4\lambda}\left(F_{c,0}^{n+1} + F_{c,1}^{n+1} \right)$$
$$+ \frac{\Delta x \Delta y}{4}\left[4 \sum_{j=j_a}^{j_b} \frac{\Delta t K}{\Delta x}\left(u_0^{c,n+1} + u_0^{c,n} - u_{N_x+1,j}^{n+1} - u_{N_x+1,j}^n \right) \right] = 0$$

We thus finally obtain the mass-preserving transmission condition, where we finally add the source term g, as:

$$\begin{aligned}
u_0^{c,n+1} &= u_0^{c,n} + \lambda \frac{\Delta t}{\Delta x}\delta_x^1 u_0^{c,n+1} - \left(\frac{\Delta t}{\Delta x} - \frac{\Delta t}{2\lambda} \right)\left(v_0^{c,n+1} + v_1^{c,n+1} \right) \\
&\quad - K\frac{\Delta t}{\Delta x}\Delta y \sum_{j=j_a}^{j_b} \left(u_0^{c,n+1} + u_0^{c,n} - u_{N_x+1,j}^{n+1} - u_{N_x+1,j}^n \right) \\
&\quad - \frac{\Delta t}{2\lambda}\left(F_{c,0}^{n+1} + F_{c,1}^{n+1} \right) + \frac{\Delta t}{2}\left(g_0^{n+1} + g_1^{n+1} \right).
\end{aligned} \quad (46)$$

Proposition 2. *The complete numerical scheme derived for the 2D-doubly parabolic-1D-hyperbolic-parabolic model is mass-preserving—by construction—across the transmission conditions in absence of source terms.*

Remark 5. *Note that the chemoattractant equation is the same as for the 1D-doubly parabolic and 2D-doubly parabolic case. Hence, the numerical schemes (33) and (19) with boundary conditions (32) and (43) can be used.*

3.2.2. Multiple Channels

In the previous paragraphs, we connected the two-dimensional domain Ω_l with a single one-dimensional channel C at $(L_x, y) \in \Omega_l$ with $y \in [a_1, b_1]$, and j_{a_1} and j_{b_1}, the positions of the endpoints of the corridor on the numerical grid. Of course, this can be extended to more channels.

Let C_m, with $m = 1, \ldots, \mathcal{M}$ be $\mathcal{M}$ corridors, connected to the two-dimensional domain Ω_l at (L_x, y_m) with $y_m \in [a_m, b_m]$ and $a_1 > 0$, $b_m < a_{m+1}$, for $m = 1, \ldots, \mathcal{M}-1$, and

$b_M < L_y$ to avoid intersections of the corridors, with equal width $k \triangle y$, $k \in \mathbb{N}$.

3.3. Implemented Algorithm.

Before presenting the numerical tests in the next Section 4, we adapt the approximation scheme for the density u, also including the source term g, implemented to solve the problem in the 2D-1D domain. As underlined before, it is necessary to use implicit schemes to consider the presence of stiff source terms. For this reason, for the approximation of the time derivatives we use the Crank–Nicolson (CN) method on the diffusion and source term, which is a second-order implicit method and the explicit central method for the convection term. Moreover, since CN is only A-stable but not L-stable [38], we also need to choose a Δt that is small enough to avoid spurious oscillations of the solution during transience.

Because of the explicit term, we have numerical restrictions on the mesh grid and time step. Furthermore, as discussed previously, we introduce artificial viscosity to avoid oscillations due to not suitable mesh grid size in dominant convection regime, which is often the case in chemotaxis models. Finally, the implicit-explicit numerical method used to compute the solutions for the density u in (8) inside the 2D domain Ω_l is:

$$
\begin{aligned}
u_{i,j}^{n+1} &= u_{i,j}^n + D_u \frac{\Delta t}{2} \left[\frac{\delta_x^2(u_{i,j}^n + u_{i,j}^{n+1})}{\Delta x^2} + \frac{\delta_y^2(u_{i,j}^n + u_{i,j}^{n+1})}{\Delta y^2} \right] \\
&\quad - \frac{\Delta t}{4} \left[\frac{\delta_x^0 F_{i,j}^{x,n}}{\Delta x} + \frac{\delta_y^0 F_{i,j}^{y,n}}{\Delta y} \right] + \frac{\Delta t}{2} \left(g_{i,j}^n + g_{i,j}^{n+1} \right) \\
&\quad \underbrace{- \Delta t \left[\frac{\delta_x^2 \theta_{i,j}^n}{2\Delta x} + \frac{\delta_y^2 \theta_{i,j}^n}{2\Delta y} \right]}_{\text{artificial viscosity}},
\end{aligned}
\qquad (47)
$$

with: $\theta_{i,j}^n := \chi(\varphi_{i,k}^n) u_{i,j}^n |\nabla \varphi_{i,j}^n|$ for $i = 1, \ldots, N_x$, $j = 1, \ldots, N_y$. As can be seen, the function θ used for the artificial viscosity is almost identical to f, with the exception of using the absolute value of $\nabla \varphi$. By using this, we increase artificial viscosity only where the gradient of the chemoattractant increases. This reduces the restriction on the meshgrid due to the condition induced by the cell Péclet number, [36]. The numerical transmission condition on the left of node $1L$ ($i = N_x + 1, j = j_a, \ldots, j_b$) is:

$$
\begin{aligned}
u_{N_x+1,j}^{n+1} &= u_{N_x+1,j}^n - D_u \frac{\Delta t}{\Delta x^2} \delta_x^1(u_{N_x,j}^n + u_{N_x,j}^{n+1}) \\
&\quad + D_u \frac{\Delta t}{2\Delta y^2} \delta_y^2 \left(u_{N_x+1,j}^n + u_{N_x+1,j}^{n+1} \right) \\
&\quad + \frac{\Delta t}{\Delta x} \left(F_{N_x+1,j}^{x,n} + F_{N_x,j}^{x,n} \right) - \frac{\Delta t}{\Delta y} \left(F_{N_x+1,j+1}^{y,n} - F_{N_x+1,j-1}^{y,n} \right) \\
&\quad + \frac{\Delta t}{2} \left(g_{N_x+1,j}^n + g_{N_x+1,j}^{n+1} \right) - \Delta t \left(\frac{\delta_x^2 \theta_{N_x,j}^n}{\Delta x} + \frac{\delta_y^2 \theta_{N_x+1,j}^n}{2\Delta y} \right) \\
&\quad \underbrace{+ K \frac{\Delta t}{\Delta x} \left(u_0^{c,n} - u_{N_x+1,j}^n + u_0^{c,n+1} - u_{N_x+1,j}^{n+1} \right)}_{\text{(additional term for transmission condition)}}.
\end{aligned}
\qquad (48)
$$

The role of permeability coefficient K in the positivity of (48) is discussed in Section 3.3.1.

For the external boundaries (the edges of the chamber Ω_l except at the junctions $j = j_a, \ldots, j_b$), we use:

$$\begin{cases}
u_{i,0}^{n+1} = u_{i,0}^n + D_u \frac{\Delta t}{2\Delta x^2} \delta_x^2 \left(u_{i,0}^n + u_{i,0}^{n+1}\right) + D_u \frac{\Delta t}{\Delta y^2} \delta_y^1 \left(u_{i,0}^n + u_{i,0}^{n+1}\right) \\
\quad - \frac{\Delta t}{2\Delta x} \delta_x^0 F_{i,0}^{x,n} - \frac{\Delta t}{\Delta y}\left(F_{i,0}^{y,n} + F_{i,1}^{y,n}\right) \\
\quad + \frac{\Delta t}{2}\left(g_{i,0}^n + g_{i,0}^{n+1}\right) - \Delta t \left(\frac{\delta_x^2 \theta_{i,0}^n}{2\Delta x} + \frac{\delta_y^1 \theta_{i,0}^n}{\Delta y}\right), \; i=1,\ldots,N_x, \\[4pt]
u_{i,N_y+1}^{n+1} = u_{i,N_y+1}^n + D_u \frac{\Delta t}{2\Delta x^2} \delta_x^2 \left(u_{i,N_y+1}^n + u_{i,N_y+1}^{n+1}\right) \\
\quad - D_u \frac{\Delta t}{\Delta y^2} \delta_y^1 \left(u_{i,N_y}^n + u_{i,N_y}^{n+1}\right) \\
\quad - \frac{\Delta t}{2\Delta x} \delta_x^0 F_{i,N_y+1}^{x,n} + \frac{\Delta t}{\Delta y}\left(F_{i,N_y}^{y,n} + F_{i,N_y+1}^{y,n}\right) \\
\quad + \frac{\Delta t}{2}\left(g_{i,N_y+1}^n + g_{i,N_y+1}^{n+1}\right) - \Delta t \left(\frac{\delta_x^2 \theta_{i,N_y+1}^n}{2\Delta x} + \frac{\delta_y^1 \theta_{i,N_y}^n}{-\Delta y}\right), \; i=1,\ldots,N_x, \\[4pt]
u_{0,j}^{n+1} = u_{0,j}^n + D_u \frac{\Delta t}{\Delta x^2} \delta_x^1 \left(u_{0,j}^n + u_{0,j}^{n+1}\right) + D_u \frac{\Delta t}{2\Delta y^2} \delta_y^2 \left(u_{0,j}^n + u_{0,j}^{n+1}\right) \\
\quad - \frac{\Delta t}{\Delta x}\left(F_{0,j}^{x,n} + F_{1,j}^{x,n}\right) - \frac{\Delta t}{2\Delta y} \delta_y^0 F_{0,j}^{y,n} \\
\quad + \frac{\Delta t}{2}\left(g_{0,j}^n + g_{0,j}^{n+1}\right) - \Delta t \left(\frac{\delta_x^1 \theta_{0,j}^n}{\Delta x} + \frac{\delta_y^2 \theta_{0,j}^n}{2\Delta y}\right), \; j=1,\ldots,N_y, \\[4pt]
u_{N_x+1,j}^{n+1} = u_{N_x+1,j}^n - D_u \frac{\Delta t}{\Delta x^2} \delta_x^1 \left(u_{N_x,j}^n + u_{N_x,j}^{n+1}\right) \\
\quad + D_u \frac{\Delta t}{2\Delta y^2} \delta_y^2 \left(u_{N_x+1,j}^n + u_{N_x+1,j}^{n+1}\right) \\
\quad - \frac{\Delta t}{2\Delta x} \delta_x^0 F_{i,N_y+1}^{x,n} + \frac{\Delta t}{\Delta x}\left(F_{i,N_y}^{y,n} + F_{i,N_y+1}^{y,n}\right) \\
\quad + \frac{\Delta t}{2}\left(g_{N_x+1,j}^n + g_{N_x+1,j}^{n+1}\right) - \Delta t \left(\frac{\delta_x^1 \theta_{N_x,j}^n}{-\Delta x} + \frac{\delta_y^2 \theta_{N_x+1,j}^n}{2\Delta y}\right), \; j=1,\ldots,N_y, \\
j \neq j_a, \ldots, j_b.
\end{cases} \quad (49)$$

For the corners, we use the following boundary conditions:

$$\begin{cases}
u_{0,0}^{n+1} = u_{0,0}^n + D_u \frac{\Delta t}{\Delta x^2} \delta_x^1 \left(u_{0,0}^n + u_{0,0}^{n+1}\right) + D_u \frac{\Delta t}{\Delta y^2} \delta_y^1 \left(u_{0,0}^n + u_{0,0}^{n+1}\right) \\
\quad - \frac{\Delta t}{\Delta x}\left(F_{0,0}^{x,n} + F_{1,0}^{x,n}\right) - \frac{\Delta t}{\Delta y}\left(F_{0,0}^{y,n} + F_{0,1}^{y,n}\right) \\
\quad + \frac{\Delta t}{2}\left(g_{0,0}^n + g_{0,0}^{n+1}\right) \\
\quad - \Delta t \left(\frac{\delta_x^1 \theta_{0,0}^n}{\Delta x} + \frac{\delta_y^1 \theta_{0,0}^n}{\Delta y}\right), \\[4pt]
u_{N_x+1,0}^{n+1} = u_{N_x+1,0}^n - D_u \frac{\Delta t}{\Delta x^2} \delta_x^1 \left(u_{N_x,0}^n + u_{N_x,0}^{n+1}\right) \\
\quad + D_u \frac{\Delta t}{\Delta y^2} \delta_y^1 \left(u_{N_x+1,0}^n + u_{N_x+1,0}^{n+1}\right) \\
\quad - \frac{\Delta t}{\Delta x}\left(F_{N_x+1,0}^{x,n} + F_{N_x,0}^{x,n}\right) - \frac{\Delta t}{\Delta y}\left(F_{N_x+1,0}^{y,n} + F_{N_x+1,1}^{y,n}\right) \\
\quad + \frac{\Delta t}{2}\left(g_{N_x+1,0}^n + g_{N_x+1,0}^{n+1}\right) \\
\quad - \Delta t \left(\frac{\delta_x^1 \theta_{N_x,0}^n}{\Delta x} + \frac{\delta_y^1 \theta_{N_x+1,0}^n}{\Delta y}\right), \\[4pt]
u_{0,N_y+1}^{n+1} = u_{0,N_y+1}^n + D_u \frac{\Delta t}{\Delta x^2} \delta_x^1 \left(u_{0,N_y+1}^n + u_{0,N_y+1}^{n+1}\right) \\
\quad - D_u \frac{\Delta t}{\Delta y^2} \delta_y^1 \left(u_{0,N_y}^n + u_{0,N_y}^{n+1}\right) \\
\quad - \frac{\Delta t}{\Delta x}\left(F_{0,N_y+1}^{x,n} + F_{1,N_y+1}^{x,n}\right) - \frac{\Delta t}{\Delta y}\left(F_{0,N_y+1}^{y,n} + F_{0,N_y}^{y,n}\right) \\
\quad + \frac{\Delta t}{2}\left(g_{0,N_y+1}^n + g_{0,N_y+1}^{n+1}\right) \\
\quad - \Delta t \left(\frac{\delta_x^1 \theta_{0,N_y+1}^n}{\Delta x} + \frac{\delta_y^1 \theta_{0,N_y}^n}{\Delta y}\right), \\[4pt]
u_{N_x+1,N_y+1}^{n+1} = u_{N_x+1,N_y+1}^n - D_u \frac{\Delta t}{\Delta x^2} \delta_x^1 \left(u_{N_x,N_y+1}^n + u_{N_x,N_y+1}^{n+1}\right) \\
\quad - D_u \frac{\Delta t}{\Delta y^2} \delta_y^1 \left(u_{N_x+1,N_y}^n + u_{N_x+1,N_y}^{n+1}\right) \\
\quad - \frac{\Delta t}{\Delta x}\left(F_{N_x+1,N_y+1}^{x,n} + F_{N_x,N_y+1}^{x,n}\right) \\
\quad - \frac{\Delta t}{\Delta y}\left(F_{N_x+1,N_y+1}^{y,n} + F_{N_x+1,N_y}^{y,n}\right) \\
\quad + \frac{\Delta t}{2}\left(g_{N_x+1,N_y+1}^n + g_{N_x+1,N_y+1}^{n+1}\right) \\
\quad - \Delta t \left(\frac{\delta_x^1 \theta_{N_x,N_y+1}^n}{\Delta x} + \frac{\delta_y^1 \theta_{N_x+1,N_y}^n}{\Delta y}\right).
\end{cases} \quad (50)$$

Similarly, for the chemoattractant ϕ, we have the implicit-explicit scheme in the interior points of the 2D domain:

$$\phi_{i,j}^{n+1} = \phi_{i,j}^n + D_\phi \frac{\Delta t}{2}\left[\frac{\delta_x^2\left(\phi_{i,j}^n + \phi_{i,j}^{n+1}\right)}{\Delta x^2} + \frac{\delta_y^2\left(\phi_{i,j}^n + \phi_{i,j}^{n+1}\right)}{\Delta y^2}\right] \\ \frac{\Delta t}{2}\left(a(u_{i,j}^n + u_{i,j}^{n+1})\right) - \frac{\Delta t}{2}\left(b(\phi_{i,j}^n + \phi_{i,j}^{n+1})\right) \tag{51}$$

At the boundaries and the corners of the numerical schemes for ϕ, we use, respectively, conditions:

$$\begin{cases} \phi_{0,0}^{n+1} = \phi_{0,0}^n + D_\phi \frac{\Delta t}{\Delta x^2}\delta_x^1(\phi_{0,0}^n + \phi_{0,0}^{n+1}) + 2D_\phi \frac{\Delta t}{\Delta y^2}\delta_y^1(\phi_{0,0}^n + \phi_{0,0}^{n+1}) \\ \quad + a\Delta t(u_{0,0}^n + u_{0,0}^{n+1}) - b\Delta t(\phi_{0,0}^n + \phi_{0,0}^{n+1}), \\ \phi_{N_x+1,0}^{n+1} = \phi_{N_x+1,0}^n - D_\phi \frac{\Delta t}{\Delta x^2}\delta_x^1(\phi_{N_x+1,0}^n + \phi_{N_x+1,0}^{n+1}) \\ \quad + D_\phi \frac{\Delta t}{\Delta y^2}\delta_y^1(\phi_{N_x+1,0}^n + \phi_{N_x+1,0}^n) \\ \quad + a\Delta t(u_{N_x+1,0}^n + u_{N_x+1,0}^{n+1}) - b\Delta t(\phi_{N_x+1,0}^n + \phi_{N_x+1,0}^{n+1}), \\ \phi_{N_x+1,N_y+1}^{n+1} = \phi_{N_x+1,N_y+1}^n - D_\phi \frac{\Delta t}{\Delta x^2}\delta_x^1(\phi_{N_x,N_y+1}^n + \phi_{N_x,N_y+1}^{n+1}) \\ \quad - D_\phi \frac{\Delta t}{\Delta y^2}\delta_x^1(\phi_{N_x+1,N_y+1}^n + \phi_{N_x+1,N_y}^n) \\ \quad + a\Delta t u_{N_x+1,N_y+1}^n - b\Delta t \phi_{N_x+1,N_y+1}^n, \\ \phi_{0,N_y+1}^{n+1} = \phi_{0,0}^n + D_\phi \frac{\Delta t}{\Delta x^2}\delta_x^1(\phi_{1,N_y+1}^n + \phi_{1,N_y+1}^{n+1}) \\ \quad - D_\phi \frac{\Delta t}{\Delta y^2}\delta_y^1(\phi_{0,N_y}^n + \phi_{0,N_y}^{n+1}) + a\Delta t u_{0,N_y+1}^n - b\Delta t \phi_{0,N_y+1}^n, \end{cases} \tag{52}$$

For the borders, we use:

$$\begin{cases} \phi_{i,0}^{n+1} = \phi_{i,0}^n + D_\phi \frac{\Delta t}{\Delta x^2}\delta_x^2(\phi_{i,0}^n + \phi_{i,0}^{n+1}) + 2D_\phi \frac{\Delta t}{\Delta y^2}\delta_y^1(\phi_{i,0}^n + \phi_{i,0}^{n+1}) \\ \quad + a\Delta t u_{i,0}^n - b\Delta t \phi_{i,0}^n, \\ \phi_{i,N_y+1}^{n+1} = \phi_{i,N_y+1}^n + D_\phi \frac{\Delta t}{\Delta x^2}\delta_x^2(\phi_{i,N_y+1}^n + \phi_{i,N_y+1}^{n+1}) \\ \quad - 2D_\phi \frac{\Delta t}{\Delta y^2}\delta_y^1(\phi_{i,N_y}^n + \phi_{i,N_y}^{n+1}) + a\Delta t u_{i,N_y+1}^n - b\Delta t \phi_{i,N_y+1}^n, \\ \phi_{0,j}^{n+1} = \phi_{0,j}^n + D_\phi \frac{\Delta t}{\Delta y^2}\delta_y^2(\phi_{0,j}^n + \phi_{0,j}^{n+1}) + 2D_\phi \frac{\Delta t}{\Delta x^2}\delta_x^1(\phi_{0,j}^n + \phi_{0,j}^{n+1}) \\ \quad + a\Delta t u_{0,j}^n - b\Delta t \phi_{0,j}^n, \\ \phi_{N_x+1,j}^{n+1} = \phi_{N_x+1,j}^n + D_\phi \frac{\Delta t}{\Delta y^2}\delta_y^2(\phi_{N_x+1,j}^n + \phi_{N_x+1,j}^{n+1}) \\ \quad - 2D_\phi \frac{\Delta t}{\Delta x^2}\delta_x^1(\phi_{N_x,j}^n + \phi_{N_x,j}^{n+1}) + a\Delta t u_{N_x+1,j}^n - b\Delta t \phi_{N_x+1,j}^n. \end{cases} \tag{53}$$

Note that for $i = N_x + 1$ the last formula in (53) is applied for $j = 1, \ldots, N_y$, $j \neq j_a, \ldots, j_b$.

Remark 6. *If we consider a two-dimensional domain Ω_r connected to the right endpoint of the one-dimensional corridor C, the complete numerical scheme for the left domain Ω_l described above can be considered.*

The main difference is that the transmission conditions at the interface between the box and the channel (the left for the box Ω_l and the right for the corridor) are reversed to the left for the corridor and the right for the box Ω_r. In the numerical scheme, the change only affects the channel C, where we have transmission conditions also for $u_{N+1}^{c,n}$ (resp. $v_{N+1}^{c,n}$). The same boundary condition can be used without transmission conditions, with only the additional term derived from the KK-condition and it must be added as well for $u_0^{c,n}$ (resp. $v_0^{c,n}$).

For the computation of solutions on the one-dimensional channel C, we have two different approximations depending on the choice of the model we assign to it. If we solve the doubly parabolic problem (9), the approximation scheme used is the Crank–Nicolson scheme, as above:

$$u_i^{c,n+1} = u_i^n + D_{u_c} \frac{\Delta t}{2} \left[\frac{\delta_x^2 (u_i^{c,n} + u_i^{c,n+1})}{\Delta x^2} \right] - \Delta t \delta_x^0 F_{c,i}^n$$
$$+ \frac{\Delta t}{2} \left(g_i^n + g_i^{n+1} \right) - \Delta t \left(\frac{\delta_x^2 \theta_i^n}{2\Delta x} \right), \quad (54)$$

with the transmission condition on the left of node 1L ($i = 0$) given by:

$$u_0^{c,n+1} = u_0^{c,n} + D_{u_c} \frac{\Delta t}{\Delta x^2} \delta_x^1 \left(u_0^{c,n} + u_0^{c,n+1} \right) - \frac{\Delta t}{\Delta x} \left(F_{c,0}^n + F_{c,1}^n \right) - \Delta t \frac{\delta_x^1 \theta_0^n}{\Delta x}$$
$$- K \frac{\Delta t}{\Delta x} \left\{ (b-a)(u_0^{c,n} + u_0^{c,n+1}) + \sum_{j=j_a}^{j_b} \left(u_{N_x+1,j}^n + u_{N_x+1,j}^{n+1} \right) \right\} \quad (55)$$
$$+ \frac{\Delta t}{2} \left(g_0^n + g_0^{n+1} \right).$$

If, instead, we need to solve the hyperbolic-parabolic problem (7), an implicit version of the second-order AHO scheme is applied—see the scheme reported in (44)—in order to ensure the stability of numerical solutions in the channels.

The scheme (44) is endowed with transmission conditions (45) and (46).

3.3.1. Stability at Interfaces

Note that, in order to ensure the positivity of the quantities in the above formulas deriving from the KK conditions—i.e., (48) for the 2D domain and (55) or (45) for the 1D domain—we also need to take care of the ratio between the KK coefficient K and the space discretization steps. In particular, for (48) and (45), one needs to ensure that $K \frac{\Delta t}{\Delta x}$ and, respectively, $K \frac{\Delta t}{\Delta x} \Delta y$ is not too big in order to damp possible high oscillations produced by the term in parentheses. Similarly, in (55) we need to check that $K \frac{\Delta t}{\Delta x} (b - a)$ is small in order to prevent the growth of the negative term.

Moreover, as previously discussed, we need to check that the numerical monotonicity condition is satisfied:

$$\frac{k_1}{\left(k_2 + \varphi_{i,j}^n \right)^2} |\partial_{x,i,j}^n \varphi_{i,j}^n| \leq \sqrt{D_M} \quad (56)$$

in the computational domain in order to ensure non-negative solutions.

Now, we consider the interface between the 2D and 1D domains. If we assume $g = 0$ and $f = \chi(\phi) \nabla \phi$, the first equation in the 2D parabolic system, (8), rewrites as:

$$\partial_t u = D \Delta u - \text{div}(u \cdot f). \quad (57)$$

The transmission condition (48) reads as:

$$u_{N_x+1,j}^{n+1} = u_{N_x+1,j}^n - D_u \frac{\Delta t}{\Delta x^2} \delta_x^1 (u_{N_x,j}^n + u_{N_x,j}^{n+1}) + D_u \frac{\Delta t}{2\Delta y^2} \delta_y^2 \left(u_{N_x+1,j}^n + u_{N_x+1,j}^{n+1} \right)$$
$$+ \frac{\Delta t}{\Delta x} \left(f_{N_x,j}^{x,n} u_{N_x,j}^n + f_{N_x+1,j}^{x,n} u_{N_x+1,j}^n \right)$$
$$- \frac{\Delta t}{2\Delta y} \left(f_{N_x+1,j+1}^{y,n} u_{N_x+1,j+1}^n - f_{N_x+1,j-1}^{y,n} u_{N_x+1,j-1}^n \right)$$
$$+ \frac{\Delta t}{\Delta x} \left(|f_{N_x,j}^{x,n}| u_{N_x,j}^n - |f_{N_x+1,j}^{x,n}| u_{N_x+1,j}^n \right) \quad (58)$$
$$+ \frac{\Delta t}{2\Delta y} \left(|f_{N_x+1,j+1}^{y,n}| u_{N_x+1,j+1}^n - 2|f_{N_x+1,j}^{y,n}| u_{N_x+1,j}^n \right.$$
$$\left. + |f_{N_x+1,j-1}^{y,n}| u_{N_x+1,j-1}^n \right)$$
$$+ \underbrace{\frac{\Delta x}{\Delta t} K \left(\left(u_0^{c,n} - v_{N_x+1,j}^n \right) + \left(u_0^{c,n+1} - u_{N_x+1,j}^{n+1} \right) \right)}_{\text{KK-transmission term}}.$$

Then, the transmission condition (55) for $g = 0$ and $f = \chi(\phi)\phi_x$ now reads as:

$$u_0^{c,n+1} = u_0^{c,n} + D_{u_c}\frac{\Delta t}{\Delta x^2}\delta_x^1\left(u_0^{c,n} + u_0^{c,n+1}\right) - \frac{\Delta t}{\Delta x}\left(u_0^{c,n}f_0^n + u_1^{c,n}f_1^n\right)$$
$$+ \frac{\Delta t}{\Delta x}\left(u_1^{c,n}|f_1^n| - u_0^{c,n}|f_0^n|\right)$$
$$+ \underbrace{\frac{\Delta x}{\Delta t}K\left(\sum_{j=j_a}^{j_b}\left(u_{N_x+1,j}^n + u_{N_x+1,j}^{n+1}\right) - (b-a)\left(u_0^n + u_0^{n+1}\right)\right)}_{\text{KK-transmission term}} \quad (59)$$

for $j = j_a,\ldots,j_b$. For the transmission condition (58), we see that monotonicity is preserved when:

$$1 - D_u\frac{\Delta t}{\Delta x^2} - D_u\frac{\Delta t}{\Delta y^2} + \frac{\Delta t}{\Delta x}f_{N_x+1,j}^{x,n} - \frac{\Delta t}{\Delta x}|f_{N_x+1,j}^{x,n}| - \frac{\Delta t}{\Delta y}|f_{N_y+1,j}^{y,n}| - \frac{\Delta t}{\Delta x}K > 0, \quad (60)$$

which gives us the stability condition for the left side of the interface.

For the right side of the interface, if we assign the doubly parabolic 1D system (9), we have the stability condition:

$$1 - D_{u_c}\frac{\Delta t}{\Delta x^2} + \lambda f_{c,0}^n - \frac{\Delta t}{\Delta x}|f_0^n| - \frac{\Delta t}{\Delta x}K(b-a). \quad (61)$$

We underline that (61) is not only influenced by the KK-constant K but also by the channel width $\sigma := (b-a)$, which must be taken care of accordingly.

Analogously, we conduct the derivation of stability conditions for the transmission conditions in the case where we assign the 2D parabolic model (57) on the left part of the interface and the hyperbolic system (10) on the right. For the sake of clarity, we rewrite the hyperbolic part of the system (10) for the density flux of individuals when we assume $g = 0$:

$$\begin{cases} \partial_t u_c + \partial_x v_c &= 0, \\ \partial_t v_c + \lambda_c^2 \partial_x u_c &= F_c. \end{cases} \quad (62)$$

The KK-transmission explicit version of conditions (46) and (45) in this case read as:

$$u_0^{c,n+1} = u_0^{c,n} + \lambda\frac{\Delta t}{\Delta x}\left(u_1^{c,n} - u_0^{c,n}\right) - \left(\frac{\Delta t}{\Delta x} - \frac{\Delta t}{2\lambda}\right)\left(v_0^{c,n} + v_1^{c,n}\right) - \frac{\Delta t}{2\lambda}\left(F_{c,0}^n + F_{c,1}^n\right)$$
$$-K\frac{\Delta t}{\Delta x}\Delta y\sum_{j=j_a}^{j_b}\left(u_0^{c,n} - u_{N_x+1,j}^n + u_0^{c,n+1} - u_{N_x+1,j}^{n+1}\right),$$
$$u_{N+1}^{c,n+1} = u_{N+1}^{c,n} + \lambda\frac{\Delta t}{\Delta x}\left(u_N^{c,n} - u_{N+1}^{c,n}\right) + \left(\frac{\Delta t}{\Delta x} - \frac{\Delta t}{2\lambda}\right)\left(v_N^{c,n} + v_{N+1}^{c,n}\right) + \frac{\Delta t}{2\lambda}\left(F_{c,N}^n + F_{c,N+1}^n\right)$$
$$-K\frac{\Delta t}{\Delta x}\Delta y\sum_{j=j_a}^{j_b}\left(u_{N+1}^{c,n} - u_{0,j}^{c,n} + u_{N+1}^{c,n+1} - u_{0,j}^{c,n+1}\right),$$
$$v_0^{c,n} = -K(b-a)u_0^{c,n} + K\Delta y\sum_{j=j_a}^{j_b}\left(u_{N_x+1,j}^{l,n} + u_{N_x+1,j}^{l,n+1}\right),$$
$$v_{N+1}^{c,n} = -K(b-a)u_{N+1}^{c,n} + K\Delta y\sum_{j=j_a}^{j_b}\left(u_{N_x+1,j}^{r,n} + u_{N_x+1,j}^{r,n+1}\right). \quad (63)$$

In order to establish the monotonicity condition, as in [39], we need to diagonalize the boundary conditions above into the diagonal variables $w_0^{n+1}, z_0^{n+1}, w_{N+1}^{n+1}, z_{N+1}^{n+1}$ with the relation $u = w + z$ and $v = \lambda(z - w)$.

We have:

$$v_0^{c,n+1} = \lambda\left(z_0^{c,n+1} - w_0^{c,n+1}\right) = -K(b-a)(z_0^{c,n+1} + w_0^{c,n+1})$$
$$+ K\Delta y\sum_{j=j_a}^{j_b}\left(u_{N_x+1,j}^{l,n+2} + u_{N_x+1,j}^{l,n+1}\right).$$

If we set $\rho := \frac{\lambda-(b-a)K}{\lambda+(b-a)K}$, $\varsigma^{n+1} := +\frac{K}{\lambda+(b-a)K} \sum_{j=j_a}^{j_b} \left(u_{N_x+1,j}^{l,n+2} + u_{N_x+1,j}^{l,n+1} \right)$ we find

$$z_0^{n+1} = \rho w_0^{n+1} + \triangle y \varsigma^{n+1}$$

and

$$\begin{aligned}
w_0^{n+1} + \rho w_0^{n+1} + \varsigma^{n+1} &= (1+\rho)w_0^n + \varsigma^n + 2\lambda \tfrac{\Delta t}{\Delta x}\left(w_1^n - \rho w_0^n - \varsigma^n\right) \\
&\quad + \tfrac{\Delta t}{2}\left((\rho-1)w_0^n + \varsigma^n + z_1^n - w_1^n\right) - \tfrac{\Delta t}{2\lambda}\left(F_{c,0}^n + F_{c,1}^n\right) \\
&\quad - \tfrac{\Delta t}{\Delta x} K_k\left((b-a)(1+\rho)\left(w_0^n + w_0^{n+1}\right) + (b-a)(\varsigma^n + \varsigma^{n+1})\right) \\
&\quad - \sum_{j=j_{a_k}}^{j_{b_k}}\left(u_{N_x+1,j}^n + u_{N_x+1,j}^{n+1}\right).
\end{aligned}$$

Now, by applying the monotonicity condition we get the following inequality:

$$(1+\rho) - 2\lambda \tfrac{\Delta t}{\Delta x}\rho + \tfrac{\Delta t}{2}(\rho-1) - \tfrac{\Delta t}{\Delta x}K(b-a)(1+\rho) - \tfrac{\Delta t}{2\lambda}F_{c,0}^n > 0$$
$$\Leftrightarrow \Delta t \leq \frac{1+\rho}{\frac{2\lambda}{\Delta x}\rho - \frac{\rho-1}{2} + K\frac{(b-a)(1+\rho)}{\Delta x} + \frac{F_{c,0}^n}{2\lambda}} \quad (64)$$

For the implicit AHO, the condition above reads as:

$$\Delta t \leq \frac{1+\rho}{K\frac{(b-a)(1+\rho)}{\Delta x} + \frac{F_{c,0}^n}{2\lambda}}.$$

In Figure 3, the time step restriction (64) is depicted for a qualitative understanding of the effect of the Kedem–Katchalsky constant K and of the channel width σ on the time step Δt. As expected, the time step Δt must be chosen smaller when either K or σ increases. Furthermore for $K = 0$ we recover the time step restriction of the AHO^2-scheme $\Delta t \leq \frac{4\Delta x}{\Delta x + 4\lambda} = 2 \cdot 10^{-3}$. Since the values of K are typically of similar magnitude to the diffusion coefficients, the additional stability restrictions caused by the hyperbolic part of the transmission conditions are minimal.

Finally, we also point out that the time step restriction for the transmission condition for the one-dimensional parabolic Equation (61) is much more severe than for the one-dimensional hyperbolic (64), which can also be seen qualitatively in the steepness of Figures 3 and 4.

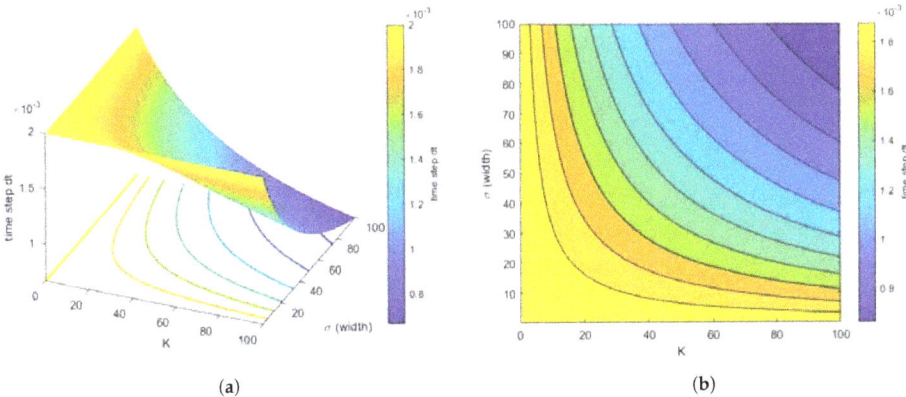

Figure 3. Time step restriction (64) Δt for the hyperbolic transmission condition with $\Delta x = 0.01$, $\Delta y = 0.1$ and $\lambda = 5$ for different K and channel widths $(b-a)$ for the transmission between the two-dimensional parabolic Equation (57) with the one-dimensional hyperbolic Equation (62) with $f = 0$.

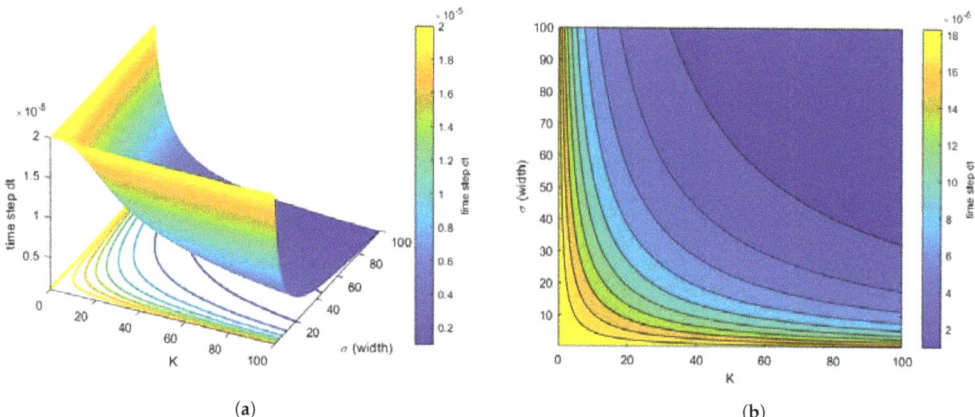

Figure 4. Time step restriction (61) $\triangle t$ for the one-dimensional parabolic transmission condition with $\triangle x = 0.01$, $\triangle y = 0.1$ and $D = 5$ for different K and channel width $\sigma = b - a$, $f = 0$. As expected, the time step $\triangle t$ must be chosen smaller when either K or the channel width σ increases.

Comparing all time step restrictions (61) and (64) with each other, it is evident that the restriction for the two-dimensional parabolic transmission condition dominates the full model.

For the sake of completeness, we underline that at each time step a non-linear equation system must be solved, for which Newton–Krylov subspace methods [40] can be used, which take advantage of the mostly sparse structure of the Jacobian matrix.

4. Numerical Tests and Results

We start this section with a preliminary test on mass-preserving properties at the boundaries. Indeed, in absence of source terms, the masses of cells and chemical substances are preserved. Then, in order to perform a numerical verification of this property, we consider the numerical approximation at the interface between 1D-1D models in the next numerical example.

Example 1. *In Figure 5, a comparison between the central mass-preserving (24) and standard finite difference (23) boundary conditions is depicted, for the 1D-doubly parabolic case on both sides of the interface (on the left) and for the 1D-doubly parabolic-1D-hyperbolic-parabolic interface (on the right). From this 1D numerical example, the necessity of developing modified boundary conditions which are consistent and preserve the mass correctly is evident. We also underline that for the discretization of the Neumann condition, the forward scheme is first-order accurate, while the central scheme is second-order accurate; for more details, see [41].*

From now on, this section is devoted to the presentation of the numerical tests and the parameters of the problem are reported in Table 1. Our aim is to show the ability of the simulation algorithm based on the model (3)–(7) to reproduce the qualitative behavior of the two population sharing the same habitat as observed in the videos of laboratory experiments, [1,2,25].

We remark that we decided to perform numerical simulations of the chip geometry by assigning the 1D-hyperbolic-parabolic model on channels, since it seems more realistic.

Table 1. Parameters of the problem.

Parameter	Description	Units	Value	Ref.
D_M	Diffusivity of cells	$\mu m^2/s$	9×10^2	[33]
D_T	Diffusivity of T-scenario 2	$\mu m^2/s$	56×10^1	empirical
D_T	Diffusivity of T-scenario 1	$\mu m^2/s$	5.6×10^1	[33]
D_φ, D_ω	Diffusivity of chemoattractants	$\mu m^2/s$	2×10^2	[33]
α_φ	growth rate of φ in scenario 2	s^{-1}/cell	0	empirical
α_φ	growth rate of φ for scenario 1	s^{-1}/cell	0.5×10^{-1}	[42]
β_φ	consumption rate of φ	s^{-1}	10^{-4}	[42]
α_ω	growth rate of ω for scenario 1	s^{-1}/cell	10^{-1}	[42]
α_ω	growth rate of ω for scenario 2	s^{-1}/cell	0	empirical
β_ω	consumption rate of ω	s^{-1}	10^{-4}	[42]
k_1	cellular drift velocity	mol cm^{2}s^{-1}	3.9×10^{-9}	[33]
k_2	receptor dissociation constant	mol	5×10^{-6}	[33]
S	maximum secretion rate of the chemicals	g/μm^3	1.9676×10^{-11}	[33]
γ	equivalent Michaelis constant	cells/μm^3	10^{-4}	[33]
L	length of the corridor	μm	500	datum
L_x	horizontal size of the box	μm	100	datum
L_y	vertical size of the box	μm	1000	datum
K	Kedem–Katchalsky constant	-	500	empirical

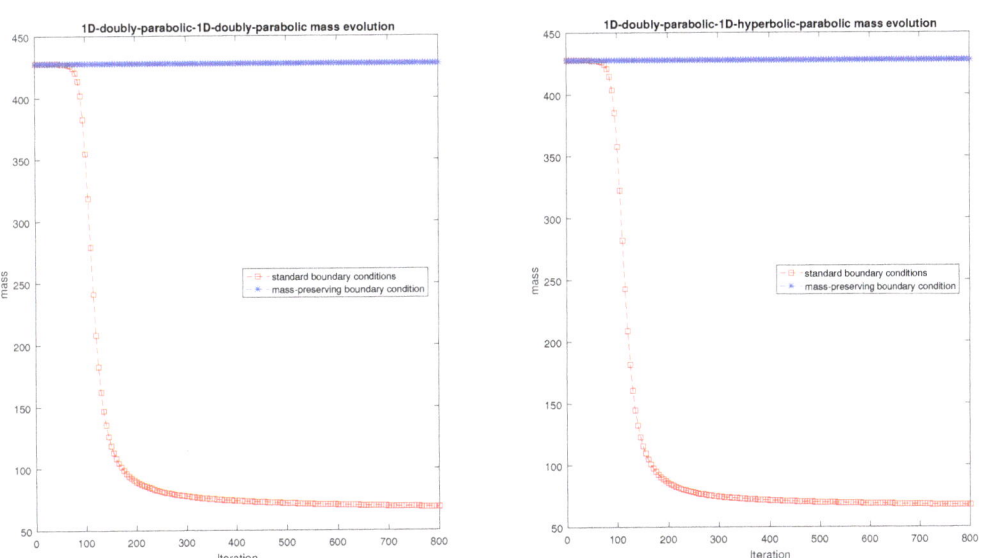

Figure 5. (**left**) evolution of total mass for 1D-1D-doubly parabolic model with standard vs. mass-preserving boundary conditions. (**right**) evolution of total mass for 1D-1D-hyperbolic-parabolic model with standard vs. mass-preserving boundary conditions.

Example 2. *Before we numerically simulate the laboratory experiment with the algorithm, we conduct a simple numerical test in order to prove its accuracy. We assumed the following setting: a left squared chamber Ω_l with one corridor positioned in the middle and only one cell family with initial distribution $u(x, y, 0) = 5e^{-\frac{1}{2}((x-0.5)^2+(y-0.5)^2)}$. Since we do not have any analytical solution for this problem, we choose Δt and $\Delta x = \Delta y$, which are small enough to obtain reasonable error estimations. In this case, we use $dt = 10^{-4}$ and $\Delta x = \Delta y = 5 \times 10^{-4}$ for the approximation u_e at time $t = 100$ and calculate the error as the quantity $\|u_e - u_{approx}\|$ in L^1-norm.*

In order to confirm the order of our scheme, we use a log-log-plot with constant and small enough Δt (resp. Δx) and decreasing Δx (resp. Δt). As shown in Figures 6 and 7, the time order and space order are equal to line with slope 2 in the log-log plot, which corresponds to our scheme of order 2 in time and space.

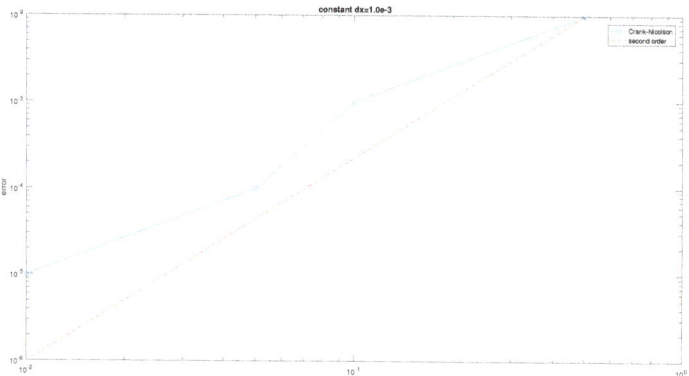

Figure 6. Log-log plot of the error—namely, the quantity $\|u_e - u_{approx}\|$ in L^1-norm as a function of the space step, with fixed $\Delta t = 10^{-3}$ and decreasing $\Delta x = 0.5, 0.1, 0.05, 0.001$ at time $t = 100$. We depict in blue the obtained error and in red a line with slope 2 for comparison.

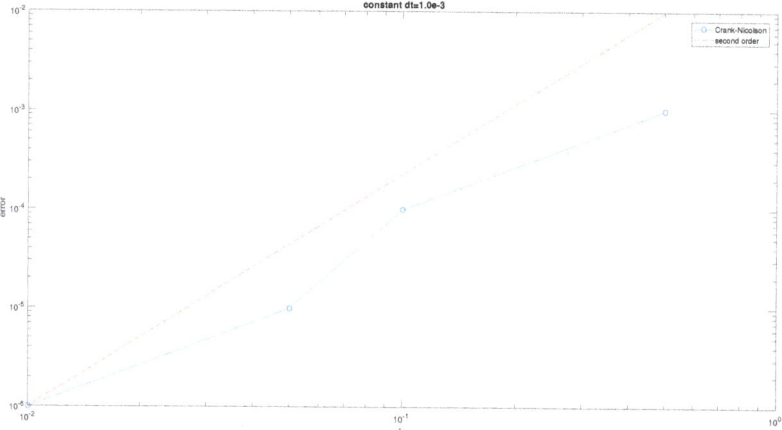

Figure 7. Log-log plot of the error, namely the quantity $\|u_e - u_{approx}\|$ in L^1-norm as a function of the time step, with fixed $\Delta x = 10^{-3}$ and decreasing $\Delta t = 0.5, 0.1, 0.05, 0.001$ at time $t = 100$. We depict in blue the obtained error and in red a line with slope 2 for comparison.

Now, we describe the simulation of the chip environment. All the simulations were performed in MATLAB©. The computational time for a simulation on the complete geometry until time $t = 50,000$ took about 400 s on an Intel(R) Core(TM) i7-3630 QM CPU 2.4 GHz. The computational domain is schematized in Figure 2 with the two chambers and 5 corridors $C_m := [0, L]$, $m = 1, \ldots, 5$ with the same width $= 12$ μm and equispaced from each other. The numerical method implemented is listed in Section 3.3; for the 1D channels, the AHO^2-scheme (of second order) is implemented, since there we are considering the hyperbolic-parabolic model. The discretization grid has time step size $\Delta t = 100$ s and space size $\Delta x = 2.5$ μm, $\Delta y = 25$ μm.

For the examples below, we assume the following initial condition (time $t = 0$) for the tumor cells distribution on the chip for $(x, y) \in \Omega_l$:

$$T(x, y, 0) = 5e^{-10^{-4}(x^2+(y-500)^2)} + 5e^{-10^{-4}(x^2+(y-5)^2)} + 5e^{-10^{-4}(x^2+(y-1000)^2)}, \quad (65)$$

Whereas, in the corridors and the right chamber, no tumor cells are present.
For the immune cells distribution on the chip for $(x,y) \in \Omega_r$, we assign:

$$M(x,y,0) = 5, \text{ for } x, y \in \Omega_r, \tag{66}$$

to mean that macrophages are disposed in the right chamber, whereas no immune cells are present in the left chamber nor in the corridors at the beginning.

For the chemoattractants, we set a constantly null initial density for ω and φ in both the chambers and also in the channels.

Example 3. *First scenario: treated case.* For the following numerical simulation, we replicate the laboratory experiment of the first scenario (treated case) described in Section 2.1; see also [1]. All the parameters are reported in Table 1.

The results are depicted in the following Figures 8 and 9.

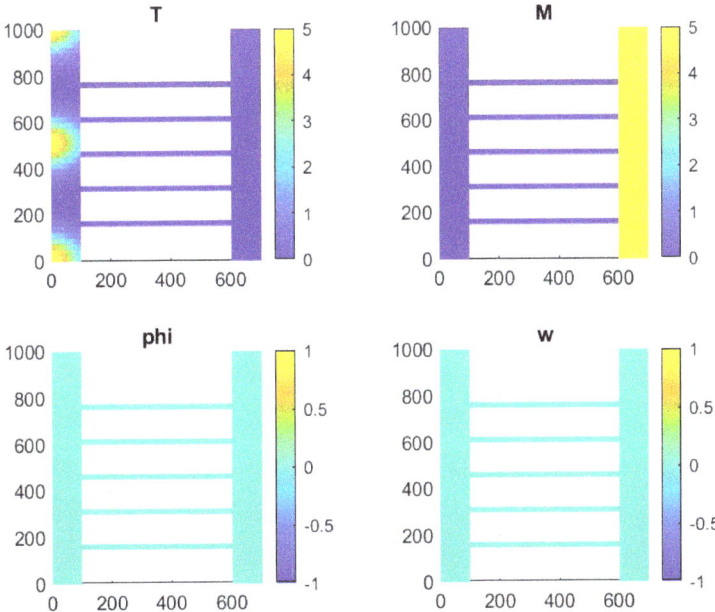

Figure 8. Treated case. Initial distribution for the model (3)–(7) at time $t = 0$.

In Figures 8 and 9, we can see the density of the tumor cells T and macrophages M for different times $t = 0$, $t = 10000$ s, and $t = 50,000$ s. Note that at time $t = 0$ tumor cells are present in the left chamber only and immune cells are present in the right chamber only.

Since tumor cells are previously treated by a chemoterapy drug, they slowly diffuse around and stay confined in the left chamber Ω_l during all the simulation time; in the meanwhile they produce chemoattractant φ attracting immune cells. Immune cells M, instead, diffuse around in Ω_r, cross the channels, and after a certain time they enter the left chamber Ω_l while creating chemoattractant ω.

This is due to the fact that the chemoattractant φ produced by cancer cells travels through channels and induces a migration of the immune cells M towards the tumor cells T, causing a higher migration towards the center of the left chamber where the initial distribution of tumor cells was closest to the chambers, as we observe from the laboratory experiment.

At the final time $t = 50,000$ s, we can see that the quantity of tumor cells is decreasing under the action of immune cells producing chemokine ω.

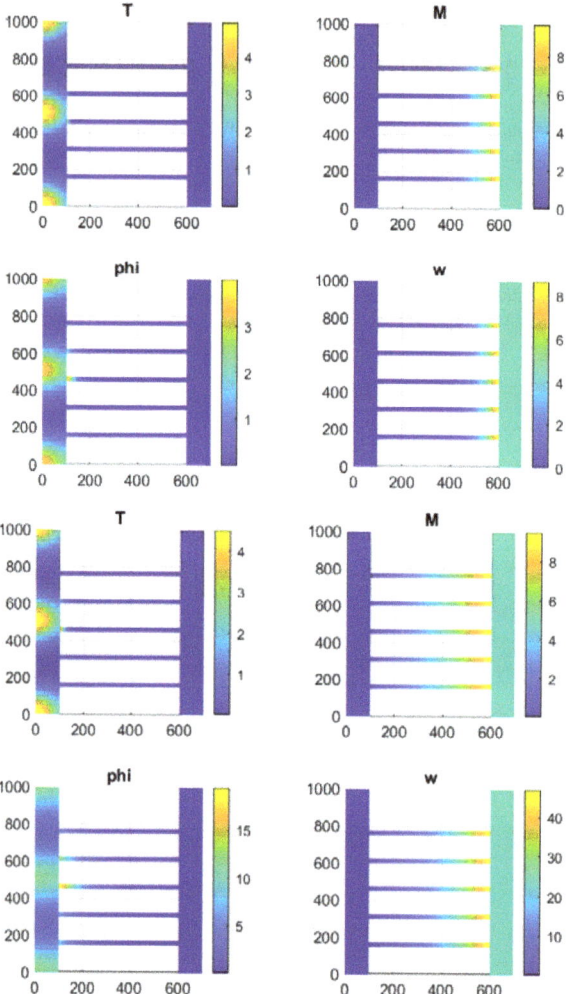

Figure 9. Treated case. Evolution of the model (3)–(7) at time $t = 10{,}000$ s (**top**) and at time $t = 50{,}000$ s (**bottom**).

Example 4. *Second scenario: untreated case.* In this numerical test, we consider the second scenario where the tumor is not treated with any medicine. Therefore, in this case we assume a higher diffusion coefficient for the tumor, but the initial conditions are the same as those used above. The results are depicted in the following Figure 10.

In Figure 10, we can see the density of the tumor cells T and macrophages M for times $t = 10{,}000$ s and $t = 50{,}000$ s. Note that at time $t = 0$ tumor cells are present in the left chamber only and immune cells are present in the right chamber only.

Since, in the laboratory experiment, untreated tumor cells diffuse around, cross the channels, and enter the right chamber Ω_r after some time, we try to reproduce such behavior by using a diffusion coefficient D_T of a order of magnitude higher respect to the one used in the first scenario. Moreover, in this case the tumor cells do not produce chemoattractant φ; for this reason, here we set $\alpha_\varphi = 0$. Immune cells M diffuse around in Ω_r and do not cross the channels, since the chemical stimulus is not secreted by T cells. Moreover, the production of the chemical substance w is neglected in this case, thus tumor cells are not killed.

We only mention that we tested the 1D-doubly parabolic model on channels and compared it with the hyperbolic-parabolic model used in the previous examples. By using the same initial data as for the other examples, we notice that the doubly parabolic model seems to have a similar pattern as for the hyperbolic-parabolic model depicted above, but the scale of the quantities differs a lot between these models.

In particular, for the doubly parabolic model the concentration of the tumor cells T is two or three order of magnitude higher. This is due to the much slower movement of the immune cells through the channels. This also explains the much higher concentration of the chemoattractant ϕ because of the much higher concentration of T compared to in the other models.

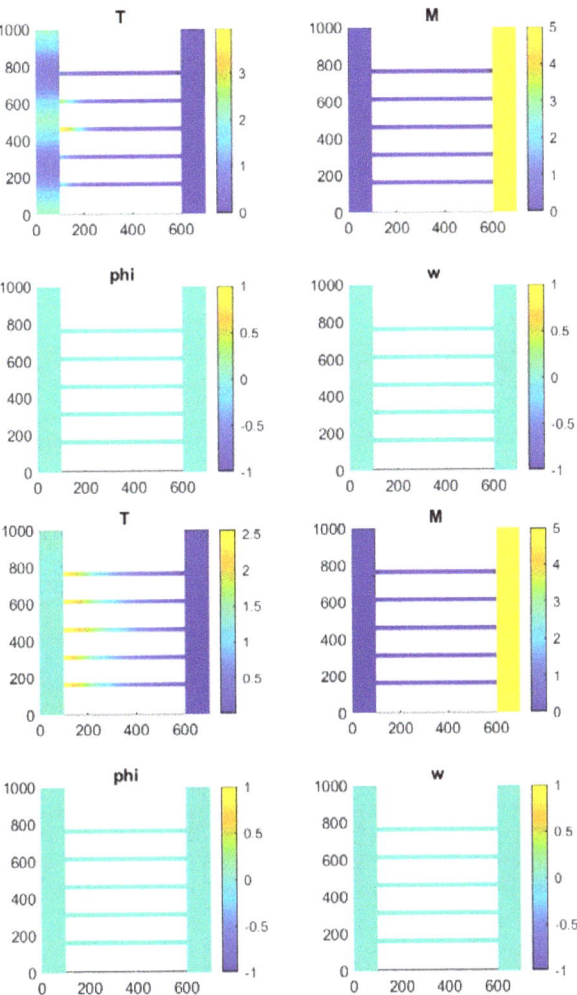

Figure 10. Untreated case. Evolution of the model (3)–(7) at time $t = 10{,}000$ s (**top**) and at time $t = 50{,}000$ s (**bottom**).

In the following Figure 11, we represent the density of tumor cells and immune cells as particles, by randomly placing them according to their density. The higher the density at a given point, the more cells will be distributed randomly around that area. If the density is lower than a chosen threshold in a certain point, no cells will be represented around it.

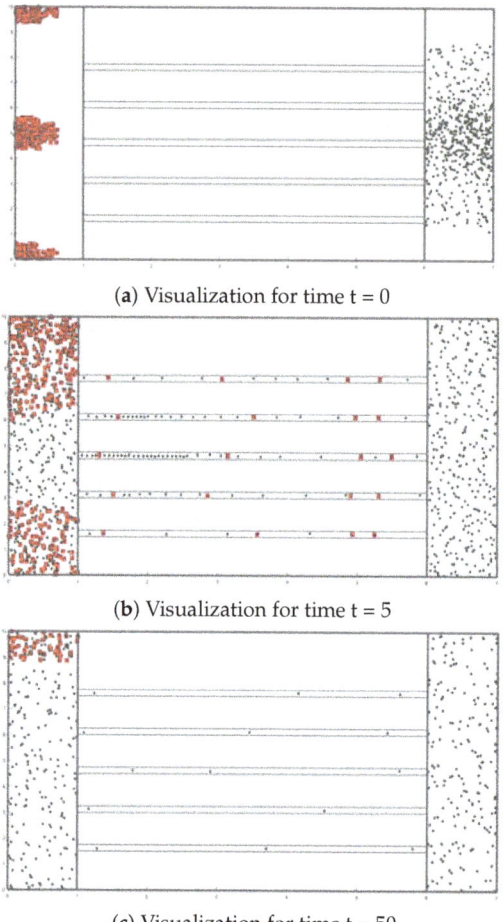

Figure 11. Visualization of immune cells (blue dots) and tumor cells (red squares) for times t = 0, t = 5, and t = 50 using the density of each quantity and representing them as cells.

5. Conclusions and Future Perspectives

The principal feature of the present work has been the development of a simulation tool to describe cell movements and interactions inside a microfluidic chip environment. Our study focused on both the modelling and the numerical point of view. Indeed, schematizing the chip geometry as two 2D boxes connected by a network of 1D channels, the main issues were:

- the introduction of mass-preserving conditions involving the balancing of incoming and outgoing fluxes passing through interfaces between 2D and 1D domains;
- the development of mass-preserving numerical schemes at the boundaries of the 2D domain and the mass-preserving transmission conditions at the 2D–1D interfaces.

Furthermore, from the modelling point of view, we studied the dynamics in the channels in the case of a doubly parabolic model and a hyperbolic-parabolic model. Since we obtained comparable asymptotic states, we decided to apply the hyperbolic-parabolic model in order to obtain a finite speed of propagation in the channels, which seems to be more realistic. In this framework, bearing in mind the laboratory experiments on a chip described in Section 2.1, it was possible to simulate the chip environment with two species of living cell moving in it.

Moreover, we remark that we can simulate more complicated situations where more than two cell species are present.

As a further development of the present study, we will work on the calibration of the model against experimental data obtained from cell tracking in a microfluidic chip [1,2,25].

Author Contributions: Methodology, G.B. and R.N.; Software, E.C.B.; Supervision, G.B. and R.N.; Validation, E.C.B. and G.B.; Visualization, E.C.B.; Data curation, E.C.B. and G.B.; Conceptualization, R.N.; Investigation, E.C.B. and G.B. Writing, E.C.B. and G.B. All authors have read and agreed to the published version of the manuscript.

Funding: This research received no external funding.

Institutional Review Board Statement: Not applicable.

Informed Consent Statement: Not applicable.

Data Availability Statement: All data are contained within the article.

Conflicts of Interest: The authors declare no conflict of interest.

References

1. Vacchelli, E.; Ma, Y.; Baracco, E.E.; Sistigu, A.; Enot, D.P.; Pietrocola, F.; Yang, H.; Adjemian, S.; Chaba, K.; Semeraro, M.; et al. Chemotherapy-induced antitumor immunity requires formyl peptide receptor 1. *Science* **2015**, *350*, 972–978. [CrossRef]
2. Businaro, L.; De Ninno, A.; Schiavoni, G.; Lucarini, V.; Ciasca, G.; Gerardino, A.; Belardelli, F.; Gabriele, L.; Mattei, F. Cross talk between cancer and immune cells: Exploring complex dynamics in a microfluidic environment. *Lab. Chip* **2013**, *13*, 229–239. [CrossRef]
3. Parlato, S.; De Ninno, A.; Molfetta, R.; Toschi, E.; Salerno, D.; Mencattini, A.; Romagnoli, G.; Fragale, A.; Roccazzello, L.; Buoncervello, M.; et al. 3D Microfluidic model for evaluating immunotherapy efficacy by tracking dendritic cell behaviour toward tumor cells. *Sci. Rep.* **2017**, *7*, 1–16. [CrossRef] [PubMed]
4. Keller, E.F.; Segel, L.A. Initiation of slime mold aggregation viewed as an instability. *J. Theor. Biol.* **1970**, *26*, 399–415. [CrossRef]
5. Greenberg, J.M.; Alt, W. Stability results for a diffusion equation with functional drift approximating a chemotaxis model. *Trans. Amer. Math. Soc.* **1987**, *300*, 235–258. [CrossRef]
6. Di Russo, C. Analysis and Numerical Approximation of Hydrodynamical Models of Biological Movements. Ph.D. Thesis, Roma Tre University (Università degli studi Roma Tre), Rome, Italy, 2011.
7. Dolak, Y.; Hillen, T. Cattaneo models for chemosensitive movement. Numerical solution and pattern formation. *J. Math. Biol.* **2003**, *46*, 153–170; Corrected Version after misprinted p.160 in *J. Math. Biol.* **2003**, *46*, 461–478. [CrossRef]
8. Filbet, F.; Laurençot, P.; Perthame, B. Derivation of hyperbolic models for chemosensitive movement. *J. Math. Biol.* **2005**, *50*, 189–207. [CrossRef]
9. Gamba, A.; Ambrosi, D.; Coniglio, A.; De Candia, A.; Di Talia, S.; Giraudo, E.; Serini, G.; Preziosi, L.; Bussolino, F. Percolation, morphogenesis, and Burgers dynamics in blood vessels formation. *Phys. Rev. Lett.* **2003**, *90*, 118101.1–118101.4. [CrossRef]
10. Perthame, B. *Transport Equations in Biology, Frontiers in Mathematics*; Birkhäuser: Basel, Switzerland, 2007.
11. Serini, G.; Ambrosi, D.; Giraudo, E.; Gamba, A.; Preziosi, L.; Bussolino, F. Modeling the early stages of vascular network assembly. *Embo J.* **2003**, *22*, 1771–1779. [CrossRef]
12. Bretti, G.; Gosse, L. Diffusive limit of a two-dimensional well-balanced approximation to a kinetic model of chemotaxis. In *SN Partial Differential Equations and Applications*; Springer: Berlin, Germany, 2021.
13. Guarguaglini, F.R.; Mascia, C.; Natalini, R.; Ribot, M. Stability of constant states and qualitative behavior of solutions to a one dimensional hyperbolic model of chemotaxis. *Discret. Contin. Dyn. Syst. Ser. B* **2009**, *12*, 39–76. [CrossRef]
14. Natalini, R.; Ribot, M. An asymptotic high order mass-preserving scheme for a hyperbolic model of chemotaxis. *SIAM J. Numer. Anal.* **2012**, *50*, 883–905. [CrossRef]
15. Gosse, L. Asymptotic-preserving and well-balanced schemes for the 1D Cattaneo model of chemotaxis movement in both hyperbolic and diffusive regimes. *J. Math. Anal. Appl.* **2012**, *388*, 964–983. [CrossRef]
16. Gosse, L. Well-balanced numerical approximations display asymptotic decay toward Maxwellian distributions for a model of chemotaxis in a bounded interval. *SIAM J. Sci. Comput.* **2012**, *34*, A520–A545. [CrossRef]
17. Bretti, G.; Natalini, R. Numerical approximation of nonhomogeneous boundary conditions on networks for a hyperbolic system of chemotaxis modeling the physarum dynamics. *J. Comput. Methods Sci. Eng.* **2018**, *18*, 85–115. [CrossRef]
18. Bretti, G.; Natalini, R.; Ribot, M. A hyperbolic model of chemotaxis on a network: A numerical study. *Math. Model. Numer. Anal.* **2014**, *48*, 231–258. [CrossRef]
19. Borsche, S.; Göttlich, S.; Klar, A.; Schillen, P. The scalar Keller-Segel model on networks. *Math. Model. Methods Appl. Sci.* **2014**, *24*, 221–247. [CrossRef]
20. Kedem, O.; Katchalsky, A. Thermodynamic analysis of the permeability of biological membrane to non-electrolytes. *Biochim. et Biophysica Acta* **1958**, *27*, 229–246. [CrossRef]

21. Quarteroni, A.; Veneziani, A.; Zunino, P. Mathematical and numerical modeling of solute dynamics in blood flow and arterial walls. *SIAM J. Num. Anal.* **2002**, *39*, 1488–1511. [CrossRef]
22. Serafini, A. Mathematical Models for Intracellular Transport Phenomena. Ph.D. Thesis, "Sapienza" University of Rome 1, Rome, Italy, 2007.
23. Cangiani A.; Natalini R. A spatial model of cellular molecular trafficking including active transport along microtubules. *J. Theor. Biol.* **2010**, *267*, 614–625. [CrossRef]
24. Di Costanzo, E.; Ingangi, V.; Angelini, C.; Carfora, M.F.; Carriero, M.V.; Natalini, R. A Macroscopic Mathematical Model For Cell Migration Assays Using A Real-Time Cell Analysis. *PLoS ONE* **2016**, *11*, e0162553. [CrossRef]
25. Agliari, E.; Biselli, E.; De Ninno, A.; Schiavoni, G.; Gabriele, L.; Gerardino, A.; Mattei, F.; Barra, A.; Businaro, L. Cancer-driven dynamics of immune cells in a microfluidic environment. *Sci. Rep.* **2014**, *4*, 6639. [CrossRef]
26. Lucarini, V.; Buccione, C.; Ziccheddu, G.; Peschiaroli, F.; Sestili, P.; Puglisi, R.; Mattia, G.; Zanetti, C.; Parolini, I.; Bracci, L.; et al. Combining Type I Interferons and 5-Aza-2'-Deoxycitidine to Improve Anti-Tumor Response against Melanoma. *J. Investig. Dermatol.* **2017**, *137*, 159–169. [CrossRef] [PubMed]
27. Nguyen, M.; De Ninno, A.; Mencattini, A.; Mermet-Meillon, F.; Fornabaio, G.; Evans, S.S.; Cossutta, M.; Khira, Y.; Han, W.; Sirven, P.; et al. Dissecting Effects of Anti-cancer Drugs and Cancer-Associated Fibroblasts by On-Chip Reconstitution of Immunocompetent Tumor Microenvironments. *Cell Rep.* **2018**, *25*, 3884–3893. [CrossRef]
28. Altrock, P.M.; Liu, L.L.; Michor, F. The mathematics of cancer: Integrating quantitative models. *Nat. Rev. Cancer* **2015**, *15*, 730–745. PMID: 26597528. [CrossRef] [PubMed]
29. Emako, C.; Gayrard, C.; Buguin, A.; de Almeida, L.N.; Vauchelet, N. Traveling Pulses for a Two-Species Chemotaxis Model. *PLoS Comput. Biol.* **2016**, *12*, 1–22. [CrossRef]
30. Preziosi, L.; Tosin, A. Multiphase and Multiscale Trends in Cancer Modellings. *Math. Model Nat. Phenom.* **2009**, *4*, 1–11. [CrossRef]
31. Méhes, E.; Vicsek, T. Collective motion of cells: from experiments to models. *Integr. Biol.* **2014**, *6*, 831–854. [CrossRef]
32. Di Costanzo, E.; Natalini, R.; Preziosi, L. A hybrid mathematical model for self-organizing cell migration in the zebrafish lateral line. *J. Math Biol.* **2015**, *71*, 171–214. [CrossRef]
33. Murray, J.D. *Mathematical Biology II Spatial Models and Biomedical Applications*; Springer: Berlin/Heidelberg, Germany, 2003.
34. Lapidis, I.R.; Schiller, R. Model for the chemotactic response of a bacterial population. *Biophys. J.* **1976**, *16*, 779–789. [CrossRef]
35. Natalini, R. Convergence to equilibrium for the relaxation approximations of conservation laws. *Comm. Pure Appl. Math.* **1996**, *49*, 795–823. [CrossRef]
36. Patankar, S.V. *Numerical Heat Transfer and Fluid Flow*; McGraw-Hill: New York, NY, USA, 1996; ISBN 0-89116-522-3.
37. Crank, J.; Nicolson, P. A practical method for numerical evaluation of solutions of partial differential equations of the heat conduction type. *Proc. Camb. Phil. Soc.* **1947**, *43*, 50–67. [CrossRef]
38. Hairer, E.; Wanner, G. Solving ordinary differential equations. II, vol. 14 of Springer Series in Computational Mathematics. In *Stiff and Differential-Algebraic Problems*, 2nd revised ed.; Springer: Berlin, Germany, 2010.
39. Aregba-Driollet, D.; Natalini, R. Convergence of relaxation schemes for conservation laws. *Appl. Anal.* **1996**, *61*, 163–193 163–193. [CrossRef]
40. Knoll, D.A.; Keyes, D.E. Jacobian-Free Newton-Krylov methods: a survey of approaches and applications. *J. Comput. Phys.* **2004**, *193*, 357–397. [CrossRef]
41. Morton, K.W.; Mayers, D. *Numerical Solutions of Partial Differential Equations*; Cambridge University Press: Cambridge, UK, 2005.
42. Curk, T.; Marenduzzo, D.; Dobnikar, J. Chemotactic Sensing towards Ambient and Secreted Attractant Drives Collective Behaviour of E. coli. *PLoS ONE* **2013**, *8*, e74878. [CrossRef] [PubMed]

Article

On-Off Intermittency in a Three-Species Food Chain

Gabriele Vissio [1,*] and Antonello Provenzale [2]

[1] Institute of Geosciences and Earth Resources, National Research Council, 10125 Turin, Italy
[2] Institute of Geosciences and Earth Resources, National Research Council, 56124 Pisa, Italy; antonello.provenzale@cnr.it
* Correspondence: gabriele.vissio@igg.cnr.it

Abstract: The environment affects population dynamics through multiple drivers. Here we explore a simplified version of such influence in a three-species food chain, making use of the Hastings–Powell model. This represents an idealized resource–consumer–predator chain, or equivalently, a vegetation–host–parasitoid system. By stochastically perturbing the value of some parameters in this dynamical system, we observe dramatic modifications in the system behavior. In particular, we show the emergence of on–off intermittency, i.e., an irregular alternation between stable phases and sudden bursts in population size, which hints towards a possible conceptual description of population outbursts grounded into an environment-driven mechanism.

Keywords: on–off intermittency; dynamical systems; theoretical ecology; stochastic forcing; hastings-powell model; food chain

1. Introduction

When Batchelor and Townsend [1] observed a peculiar irregularity in a turbulent fluid, namely the alternation between sudden bursts of motion and a milder, non-turbulent activity, they used the word *intermittency* to describe it. Since then, the same term has been used to describe several types of switching behavior between different dynamical regimes. Here, we are especially interested in the phenomenon called *on–off intermittency* [2].

On-off intermittency has been observed in real systems, such as electronic circuits [3], earthquakes [4], solar cycles [5], electrodynamics of liquid crystals [6], as well as theoretically studied through numerical approaches [2,7] with specific focus on discrete-time population dynamics models (i.e., maps) [8–10].

The goal of this work is to expand these studies to the case of a stochastically driven system of coupled ordinary differential equations (ODEs). To this end, we include the random variability of suitable model parameters to simulate environmental stochasticity in a system representing the population dynamics of three different species. In the autonomous case, a three-dimensional ODE system is the minimum requirement to allow chaotic dynamics owing to the Poincaré–Bendixson theorem [11].

Our choice here is the well-known Hastings–Powell model [12,13], a system that describes the evolution of three species, anonymously called x, y, z, which represent primary producers (resource), consumers (or host) and predators (or parasitoids). We stochastically perturb some of the system parameters, which measure species interactions or carrying capacity, showing that on–off intermittent behaviour can emerge. This feature could qualitatively explain the onset of outbreaks (also called irruptions) in the population size of some of the ecosystem components [14,15].

Section 2 describes the properties of on–off intermittency, summarizing results from previous studies that inspired this work. Section 3 introduces the Hastings–Powell model, describes its parameters and explains the numerical approach that is adopted. In Section 4, we illustrate the occurrence of on–off intermittency when one introduces environmentally driven—and stochastically simulated—parameters. Finally, Section 5 summarizes our results and outlines some possible future developments.

2. On-Off Intermittency

On-off intermittency is characterized by the alternation between *regular phases*, which duration can span a rather wide range of orders of magnitude, and *burst phases*, where a sudden instability throws the system into (possibly) chaotic behavior. This kind of intermittency can appear in a dynamical system that has an invariant manifold (in the simplest case, a fixed point) whose stability properties depend on an external control parameter but whose phase-space position is only weakly dependent upon the same parameter.

When such control parameter has an irregular temporal variation, either stochastic or chaotic, the manifold alternates between stable and unstable conditions. In order to realize on–off intermittency, a system must keep its dynamics in the proximity of the manifold, which in the stable phases must be attractive enough to allow for long periods during which the system resides in the vicinity of the manifold. Lingering near this temporarily stable manifold, the system undergoes protracted regular phases, when suddenly the volatility of the control parameter induces the instability of the manifold and causes the system to burst away from it, leading to values which are quite different from its typical statistics.

In past years, on–off intermittency has raised some interest in the scientific community. After its basics were scouted by the work of Platt et al. [2], Heagy et al. [7] gave a mathematical sounding demonstration of the power law underlying the duration of laminar phases for maps with the specific form $y_{n+1} = z_n f(y_n)$ (with the variable z_n coming from a random or a chaotic process), then Toniolo et al. [8] further deepened this latter aspect, inspecting the occurrence of on–off intermittency in a stochastically driven logistic map. Due to the possibility of adopting this concept to qualitatively explain ecological outbreaks, in 2010 Metta et al. [9] and Moon [10] investigated Toniolo's framework in the context of coupled logistic equations. While the former focused on kurtosis as an index to identify on–off intermittency, the latter put the spotlight on the stability of the coupled system, employing the largest Lyapunov Exponent to quantify the chaotic dynamics occurring with different coupling strengths of adjacent logistic systems. Here, we continue the exploration of on–off intermittency in the context of ecosystem dynamics and study its presence and characteristics in a system of coupled ordinary differential equations representing a three-layer food chain.

3. Hastings–Powell Model

Alan Hastings and Thomas Powell introduced a three-dimensional dynamical system [12] in order to illustrate chaotic behavior in a food web involving three trophic levels. They employed the type 2 functional response (i.e., a Michaelis–Menten functional form) shown in the 1975 Murdoch and Oaten's paper [16] to couple the different trophic levels of the system. The basic equations of the Hastings–Powell model are:

$$\frac{dX}{dT} = R_0 X \left(1 - \frac{X}{K_0}\right) - C_1 \frac{A_1 X}{B_1 + X} Y \tag{1}$$

$$\frac{dY}{dT} = \frac{A_1 X}{B_1 + X} Y - \frac{A_2 Y}{B_2 + Y} Z - D_1 Y \tag{2}$$

$$\frac{dZ}{dT} = C_2 \frac{A_2 Y}{B_2 + Y} Z - D_2 Z \tag{3}$$

where X, Y, Z represent the biomass of three species on three different trophic levels and T is time. Throughout the three equations, subscripts 0, 1, 2 indicate parameters referring to, respectively, X, Y, Z. R_0 and K_0 are, respectively, the growth rate and the carrying capacity of the species X. The constants A_1, A_2, B_1, B_2 characterize the functional responses among the species, representing the saturation of the response; specifically, the Bs are the prey populations that correspond to half the maximum value of the predation rate per unit prey. C_1^{-1}, C_2 are the conversion rates from resource to consumer and from prey to predator, respectively, while, finally, D_1, D_2 are constant death rates.

A suitable nondimensionalization leads to redefine the variables of the system:

$$x = \frac{X}{K_0}$$
$$y = \frac{C_1 Y}{K_0}$$
$$z = \frac{C_1 Z}{C_2 K_0}$$
$$t = R_0 T$$
(4)

Consequently, the nondimensional parameters are:

$$a_1 = \frac{K_0 A_1}{R_0 B_1} \quad b_1 = \frac{K_0}{B_1} \quad d_1 = \frac{D_1}{R_0}$$
$$a_2 = \frac{C_2 A_2 K_0}{C_1 R_0 B_2} \quad b_2 = \frac{K_0}{C_1 B_2} \quad d_2 = \frac{D_2}{R_0}$$
(5)

Thus, the final equations of Hastings–Powell model are:

$$\frac{dx}{dt} = x(1-x) - \frac{a_1 x}{b_1 x + 1} y \tag{6}$$

$$\frac{dy}{dt} = \frac{a_1 x}{b_1 x + 1} y - \frac{a_2 y}{b_2 y + 1} z - d_1 y \tag{7}$$

$$\frac{dz}{dt} = \frac{a_2 y}{b_2 y + 1} z - d_2 z \tag{8}$$

Hastings and Powell chose the model parameters to be, in their words, "biologically reasonable". For example, the parameter values associated with the consumer (y) are larger than those for the predator/parasitoid (z), so that x and y interact on a faster time scale with respect to y and z. We defer to the original work of Hastings and Powell for further discussions on parameter values.

Note that Equation (8) is conceptually different from Equations (6) and (7): indeed, it is possible to factorize z on the right hand side, leading to $\frac{dz}{dt} = \left(\frac{a_2 y}{b_2 y + 1} - d_2 \right) z$. A separation of variables allows to retrieve the exact solution $z(t) = z(0) \exp\left(\frac{a_2 y}{b_2 y + 1} - d_2 \right)$, which could replace the differential equation in the numerical simulation. This peculiarity makes Equation (8) quite different from the other two equations and, therefore, we expect it to react differently to stochastic forcing.

In Appendix A we provide a concise analysis of the fixed points in the Hastings–Powell model.

Stochastic Parameters and Numerical Simulations

To simulate how the environment affects the evolution of the three-species food chain in the Hasting-Powell model, we allow some of the model parameters to become random numbers. In particular, we allow either a_1 or K_0 in Equations (5)–(8) to vary stochastically with a uniform distribution between 0 and α. The computation of the stochastic term is performed at every time step of the numerical simulation, feeding the same term throughout all the steps needed by the Runge–Kutta 4 scheme employed. The random number at each time step is independent of the previous value, that is we force the system with white noise.

The different cases are run for 10^8 time units, after a spin-up time of 10^6 time units to eliminate the initial transient. Initial conditions for x_0, y_0 are randomly and uniformly chosen between 0 and 1 while the initial value for z_0 is randomly and uniformly chosen between 4 and 5. This choice for z_0 is related to the convenience of starting as close as possible to the system attractor, thus reducing the spin-up time. Laminar phases are defined as $x > 1 - 0.001$ or $y, z < 0.001$—i.e., a distance of 10^{-3} from the stable fixed point.

4. Results

4.1. Intensity of Grazing

The parameter a_1 measures the intensity of grazing by the consumer (y) on the resource (x) in the coupling term between the equations for x and y in Equations (6) and (7). As a first test, we replace a_1 with the random number $\tilde{a}_1$, uniformly distributed between 0 and α. In this way, the time-averaged value of $\tilde{a}_1$ becomes $\bar{a}_1 = \alpha/2$. Therefore, the coupling between x and y becomes:

$$\frac{\tilde{a}_1 x}{b_1 x + 1} y \tag{9}$$

Here we use $\alpha = 3.5$, which gives $\bar{a}_1 = 1.75$. For the other parameters we adopt the same values as in the original paper of Hastings and Powell, namely:

$$b_1 = 3 \qquad d_1 = 0.4 \qquad a_2 = 0.1 \qquad b_2 = 2 \qquad d_2 = 0.01 \tag{10}$$

Figure 1 shows the time series of the three trophic levels (x, y and z) and a running mean of the instantaneous value of $\tilde{a}_1$. The time series of the resource x and of the herbivorous y visually illustrate the occurrence of on–off intermittency, with the alternation of long laminar phases and irregularly spaced bursts. The laminar phases of x are centered on $x = 1$, corresponding to a fixed point of the system, and the bursts are towards lower values when the herbivorous density suddenly increases. The z signal, instead, corresponds to a smoothed version of the intermittent signals and it is slightly delayed with respect to the herbivorous dynamics, as expected from the form of the equations.

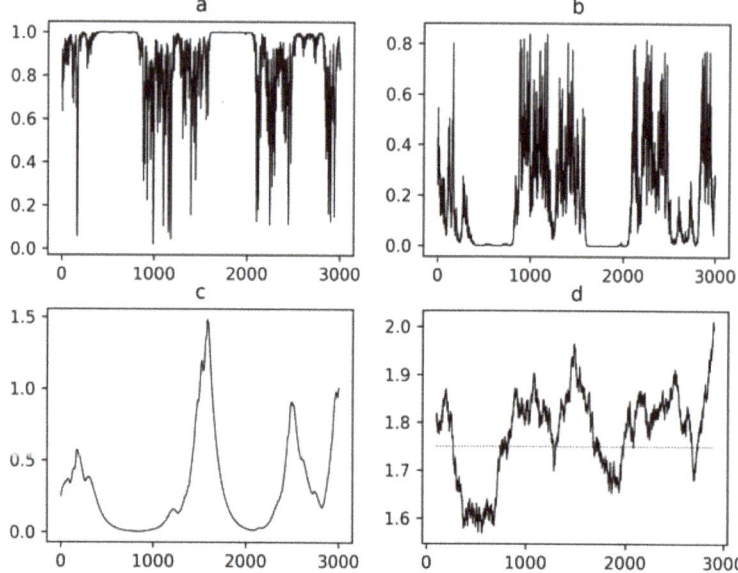

Figure 1. Case $\alpha = 3.5$ for stochastic a_1. Time series of x (Panel **a**), y (Panel **b**), z (Panel **c**) and of the running mean of $\tilde{a}_1$ computed on a window with width $\tau = 200$—the dotted line is the mean value of $\tilde{a}_1$ (Panel **d**).

As mentioned above, the simplest case of on–off intermittency appears when the stability of a fixed point of the system depends on an external parameter that varies irregularly in time, thus determining an alternation between stability and instability of the fixed point. To motivate our choice of $\alpha = 3.5$, in Figure 2 we show the orbit diagram for y in the range $0 \leq a_1 \leq 5$ (the orbit diagram for x is conceptually similar). In order to find on–off intermittency, we need to span a parameter range covering the interval between stability and chaos. The chosen value of α suits this well, forcing the instantaneous value of a_1 to vary between 0 and 3.5. From Figure 1, one sees an approximate correspondence between intermittent bursts and periods when the running average of $\tilde{a}_1$ exceeds its average value which, in this case, approximately corresponds to the stability limit of the fixed point.

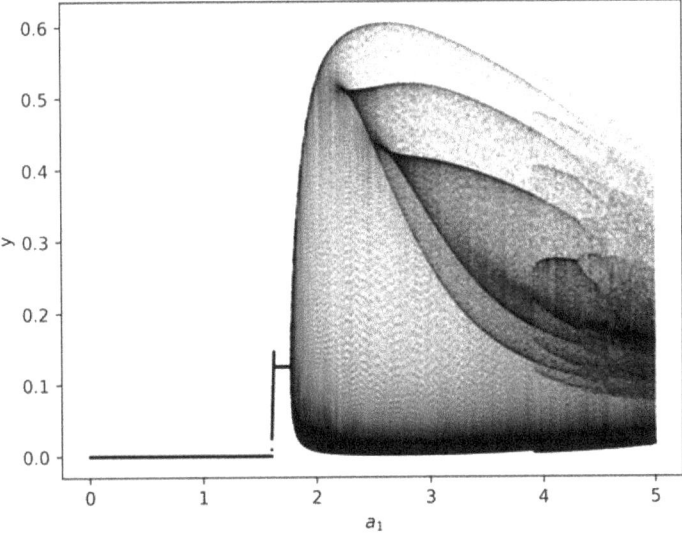

Figure 2. Orbit diagram depicting the attractors of y as a function of a_1. Other parameter values as in the original Hastings–Powell model.

The first distinctive feature of on–off intermittent time series is the shape of the probability distribution of the off-phase durations—i.e., the number of time steps in which the system endures off (laminar) behaviour. It has been shown that for a simplified type of discrete maps [7,8], for on–off intermittency the distribution of laminar phase duration, D, follows a power law, $D^{-\frac{3}{2}}$. Figure 3 shows that, also for this continuous on–off intermittent system, the x and y signals display the same approximate power-law distribution of off phases, at least in a limited range of off-phase durations.

Another characteristics of the intermittent signals is the broad distribution of the amplitudes. Figure 4 shows the distribution of maxima for on–off intermittency and for standard chaotic behavior with a fixed value $a_1 = 5$. An approximate power-law distribution of the maxima is evident for the intermittent dynamics.

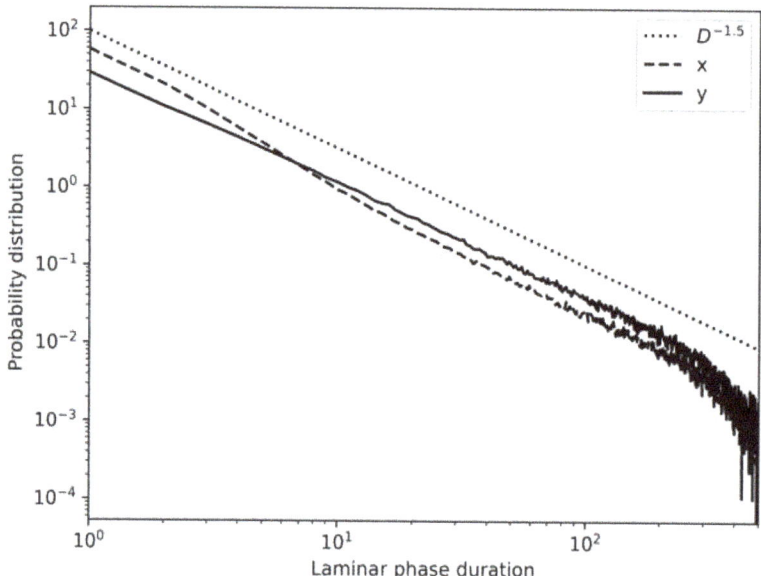

Figure 3. Duration of the off (laminar) phases of the x (dashed) and y (solid) components with a stochastic a_1 parameter in the Hastings–Powell model. The dotted line indicates a dependence proportional to $D^{-\frac{3}{2}}$.

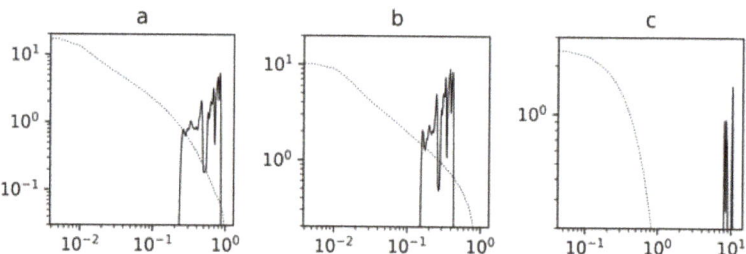

Figure 4. Probability distribution of the maxima of $1 - x$ (Panel **a**), y (Panel **b**), z (Panel **c**), in case of non-intermittent dynamics (solid line) and for on–off intermittent behavior (dotted line).

Conceptually, inserting the stochastic term as done in Equation (9) is tantamount to randomly forcing A_1 in Equations (1) and (2). Thus, the results presented in this section indicate that the environmental fluctuations (represented by the stochastic term), randomly influencing the rate of successful consumption by y of the resource x, can cause on–off intermittency in both compartments.

4.2. Carrying Capacity K_0

Another interesting option is to allow the environment to stochastically affect the system carrying capacity, K_0. Even though a_1 and K_0 are mathematically related, their ecological meaning is different: the former is related to the interaction between x and y and the latter only to the maximum value of x in the absence of consumers. Therefore, they deserve separate analyses from an ecological standpoint. Looking at Equation (5), we infer that to this end we must multiply a_1, b_1, a_2 and b_2 in Equations (6)–(8) by the same value of a random number ρ, uniformly distributed in the interval $0 \leq \rho \leq \alpha$. The parameters chosen for this Section are the same described in Equation (10), with $a_1 = 2$. Following the rationale that yielded Equation (9), the couplings in Equations (6)–(8) become:

$$\frac{\tilde{a}_1 x}{\tilde{b}_1 x + 1} y \tag{11}$$

$$\frac{\tilde{a}_2 y}{\tilde{b}_2 y + 1} z \tag{12}$$

Choosing different values for α leads to different and peculiar behaviours.

From $\alpha = 0$ to $\alpha \approx 1.3$ the system undergoes a long lasting stability at $x = 1, y = 0$, $z = 0$. For $\alpha = 1.4$ we observe the occurrence of on–off intermittency for x and y, while after the transient the z species becomes extinct, that is, the predator (parasitoid) cannot control the consumer (host).

Figure 5 shows the time series of x and y, along with the probability distributions of the maxima and the moving average of the value of the random number ρ controlling K_0. As in the case of stochastic variability in a_1, the intermittency of the time series is matched by the fluctuations of the running mean of K_0, with low values of the latter corresponding to laminar phases of the time series.

Figure 6 shows the probability distribution of the laminar phase durations, which matches a power law with $D^{-\frac{3}{2}}$.

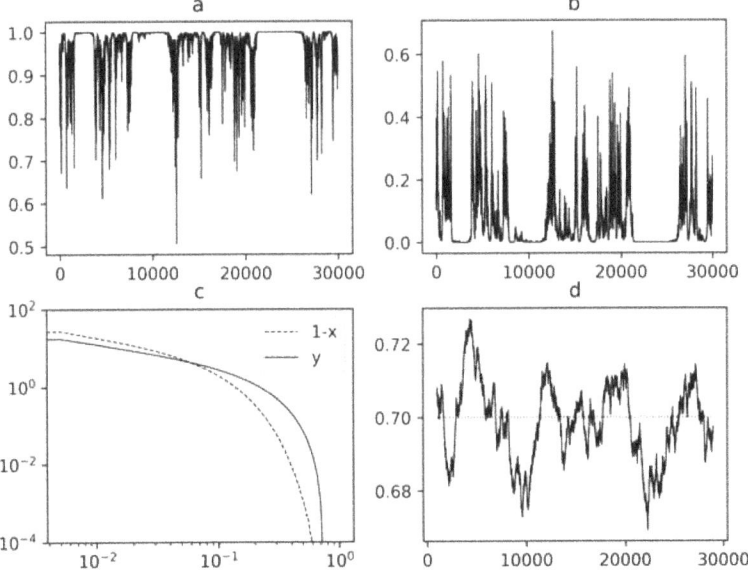

Figure 5. Stochastic K_0 with $\alpha = 1.4$. Time series of x (Panel **a**), y (Panel **b**) and of the running mean of the random variable ρ controlling K_0, computed on a window with width $\tau = 2000$—the dotted line is the mean value of ρ (Panel **d**). The probability distributions of the maxima of $1 - x$ and y are shown in (Panel **c**).

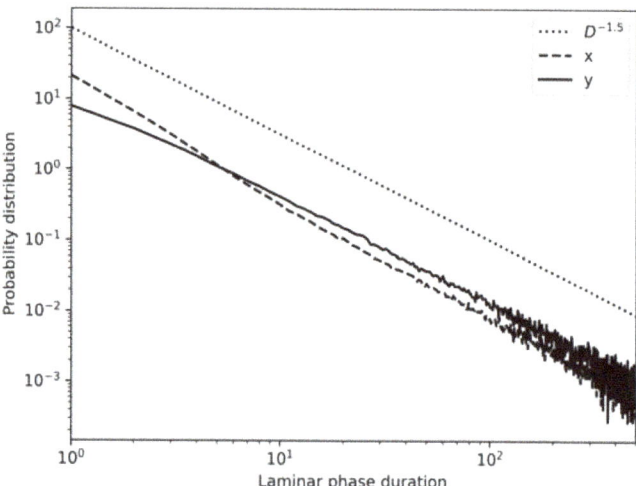

Figure 6. Laminar phase duration of the x (dashed) and y (solid) variables for stochastic variability of the carrying capacity K_0 in the case $\alpha = 1.4$. The dotted line is proportional to $D^{-\frac{3}{2}}$.

For $\alpha = 1.5$, x and y show chaotic dynamics, but the most intriguing phenomenon is related to the apparent on–off intermittency of z, as shown in Figure 7. A close inspection of the laminar phase durations, however shows that extended laminar periods are quite likely to occur, thus the curve is less steep than $D^{-\frac{3}{2}}$.

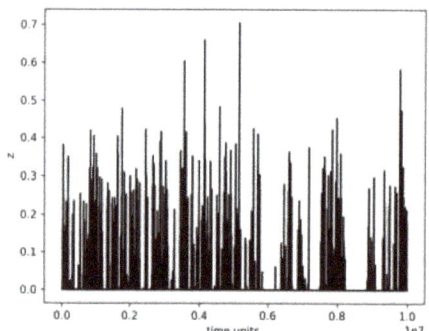
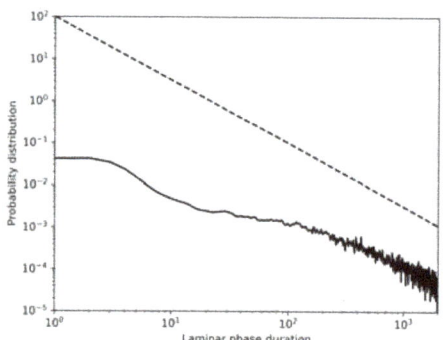

Figure 7. (**Left** panel) Time series of z in the case $\alpha = 1.5$; (**Right** panel) Laminar phase durations of the predator/parasitoid z (solid) for stochastic K_0 with $\alpha = 1.5$. The dashed line is proportional to $D^{-\frac{3}{2}}$.

Finally, we report that at larger values of α—here we employ $\alpha = 2.1$—non-intermittent, chaotic dynamics for z is paired with approximately on–off intermittent behavior for x and y. Figure 8 shows a time series of y along with the laminar phase durations, which approximately follows the simple power law (even though less robustly than in Figure 6).

The implications of these results are intriguing. If the environment induces a random variability of the carrying capacity K_0, allowing it to temporarily reach large enough values, on–off intermittency can emerge quite easily in both x and y—that is, in the primary producers and their consumers. By further increasing α, that is, the amplitude of the random variability of K_0, a peculiar intermittency in the predator (parasitoid) z develops. This implies that sudden bursts in a population could be induced far from the trophic level that is directly affected by the environmental fluctuations. For even larger values of the fluctuations in K_0 ($\alpha = 2.1$), the resource and the consumer again undergo approximate on–off behavior, while the predator behaves chaotically. Clearly, the complexity of the system behavior is huge, and a deeper exploration of the different dynamics and of their ecological implications is deferred to future works.

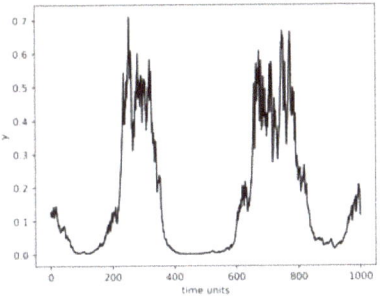 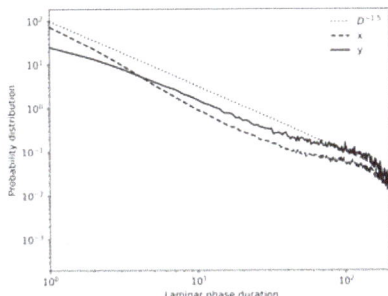

Figure 8. (**Left** panel) Time series of y in case $\alpha = 2.1$; (**Right** panel) Laminar phase durations of the x (dashed) and y (solid) variables for stochastic K_0 with $\alpha = 2.1$. The dotted line is proportional to $D^{-\frac{3}{2}}$.

5. Discussion and Conclusions

This paper conceptually extends the works of Platt et al. [2], Heagy et al. [7], Toniolo et al. [8] and Metta et al. [9] and it focuses on the emergence of on–off intermittency in idealized food chains. In our view, such dynamical behavior can be taken as a conceptual description of species outbreak events in different levels of the food chain.

To explore this issue, we used the Hastings–Powell model, a well-known system that allowed us to inspect a simple three-species food chain: resource, consumer and predator (or else, vegetation, pest host species and parasitoid). Environmental forcing was supposed to act on the resource dynamics, and it was represented as an imposed random variation in some of the controlling parameters.

When the stochastic variability is inserted into the parameter controlling the intensity of resource consumption (i.e., y on x, Section 4.1), on–off intermittency easily arises in both these variables, while the predator z displays a smoother dynamics.

Stochastic variability of the carrying capacity (Section 4.2) leads to intriguing results; indeed, increasing the range of random variations to higher values sequentially generates different behaviours:

- For low maximum values of the carrying capacity, we observe only a stable fixed point for x, y and z;
- Above a threshold of the maximum value of the carrying capacity, we observe on–off intermittency in x and y, while z goes extinct;
- For larger ranges of random variations, chaotic dynamics for x and y and intermittent behavior in z;
- For still larger fluctuations, we observe weak on–off intermittency in x and y and chaotic behavior of z.

Ecologically, this suggests that a low carrying capacity for x implies that the species directly feeding on it (y) can endure but on average it does not supply enough biomass for z to survive. A slightly larger value of K_0 allows for z to "jump start", while a higher average value of the carrying capacity is enough to fully support the predator species, bringing back on–off intermittency in the dynamics of x and y. Of course, this is just an euristic representation that requires further exploration.

Several points remain open to investigation, such as:

- Would a deterministic, chaotic system representing the environment dynamics, in place of the stochastic process adopted here, allow for a more thoroughly mathematical analysis of the problem?

- How would spatial extension, with coupling across different location of the same species, affect our results?
- How would on–off intermittency manifest itself (if it does) in a food web rather than a simple food chain?
- Can we find intermittency when using real-world datasets or controlled laboratory experiments in microcosms?

Such questions are, in our opinion, relevant to better understand bursting phenomena in ecology and will be a subject of future research, after the first demonstration of the possibility of on–off intermittent behavior in model food chains that was illustrated here.

Author Contributions: Conceptualization, G.V. and A.P.; methodology, software and validation, G.V.; writing—original draft preparation, G.V.; writing—review and editing, G.V. and A.P.; funding acquisition, A.P. All authors have read and agreed to the published version of the manuscript.

Funding: This research was partially funded by the EU H2020 project "EOTIST", Grant Agreement no. 952111.

Institutional Review Board Statement: Not applicable.

Informed Consent Statement: Not applicable.

Data Availability Statement: Code used in numerical simulation can be found at https://figshare.com/articles/software/VissioProvenzale2021/14639556 (accessed on 8 July 2021).

Acknowledgments: The authors acknowledge the three reviewers for their insightful comments. In particular, the authors are grateful to Reviewer 1, whose observations led to the material included in Appendix A.

Conflicts of Interest: The authors declare no conflict of interest.

Appendix A. Fixed Points in Hastings–Powell Model

A thorough analysis of the fixed points of the models is beyond the scope of this paper. Nevertheless, it is useful to briefly recap them, in order to give some perspective on the results obtained, especially on the population values typically attained during the laminar phases. Checking the fixed points in the system leads to four different results:

- $(0,0,0)$ Computing the Jacobian and substituting these values leads to the eigenvalues $1, -d_1, -d_2$. Note that the ecologically-relevant case has $d_1, d_2 > 0$, so this fixed point is a saddle. If we numerically perturb the system along the x direction (e.g., adding a small perturbation to $x = 0$), the system falls into the $(1,0,0)$ fixed point (see below) while, perturbing it along other directions, the null state is attractive.
- $(1,0,0)$ The three eigenvalues are -1, $\frac{a_1}{(b_1+1)} - d_1$, $-d_2$. -1 and $-d_2$ are always negative, but the second eigenvalue depends on the values of the parameters a_1, b_1. With the values used in the paper, the eigenvalue is positive and the system is therefore repulsive along one direction (it can be numerically checked perturbing $y = 0$). If $a_1 < 1.6$, then the fixed point becomes stable.
- From Equation (8), we obtain $y^* = \frac{d_2}{a_2 - d_2 b_2}$; inserting it in Equation (6) leads to two (rather cumbersome) different solutions for x^* and, consequently, two solutions for z^* from Equation (7). With the parameter values adopted in this work, one solution for x^* is negative and therefore not acceptable. The other solution is positive and, for the first case in Section 4.1 (with $a_1 = 5$, as in Hastings–Powell's paper), the fixed point is $x^* \sim 0.819$, $y^* \sim 0.125$, $z^* \sim 9.808$ (numerically verified).

References

1. Batchelor, G.; Townsend, A. The nature of turbulent motion at large wave-numbers. *Proc. R. Soc. Lond. Ser. Math. Phys. Sci.* **1949**, *199*, 238–255.
2. Platt, N.; Spiegel, E.A.; Tresser, C. On-off intermittency: A mechanism for bursting. *Phys. Rev. Lett.* **1993**, *70*, 279–282. [CrossRef] [PubMed]

3. Hammer, P.; Platt, N.; Hammel, S.; Heagy, J.; Lee, B. Experimental Observation of On-Off Intermittency. *Phys. Rev. Lett.* **1994**, *73*, 1095–1098. [CrossRef] [PubMed]
4. Bottiglieri, M.; Godano, C. On-off intermittency in earthquake occurrence. *Phys. Rev. E* **2007**, *75*, 026101. [CrossRef] [PubMed]
5. Platt, N.; Spiegel, E.A.; Tresser, C. The intermittent solar cycle. *Geophys. Astrophys. Fluid Dyn.* **1993**, *73*, 147–161. [CrossRef]
6. John, T.; Stannarius, R.; Behn, U. On-Off Intermittency in Stochastically Driven Electrohydrodynamic Convection in Nematics. *Phys. Rev. Lett.* **1999**, *83*, 749–752. [CrossRef]
7. Heagy, J.; Platt, N.; Hammel, S.M. Characterization of on-off intermittency. *Phys. Rev. E* **1994**, *49*, 1140–1150. [CrossRef] [PubMed]
8. Toniolo, C.; Provenzale, A.; Spiegel, E. Signature of on–off intermittency in measured signals. *Phys. Rev. E* **2002**, *66*, 066209. [CrossRef] [PubMed]
9. Metta, S.; Provenzale, A.; Spiegel, E. On–off intermittency and coherent bursting in stochastically-driven coupled maps. *Chaos Solitons Fractals* **2010**, *43*, 8–14. [CrossRef]
10. Moon, W. *On-Off Intermittency in Locally Coupled Maps*; Woods Hole Oceanographic Institution: Falmouth, MA, USA, 2010.
11. Strogatz, S. *Nonlinear Dynamics and Chaos: With Applications to Physics, Biology, Chemistry, and Engineering*, 2nd ed.; Westview Press: Boulder, CO, USA, 2014.
12. Hastings, A.; Powell, T. Chaos in a Three–Species Food Chain. *Ecology* **1991**, *72*, 896–903. [CrossRef]
13. Kot, M. *Elements of Mathematical Ecology*; Cambridge University Press: Cambridge, UK, 2001. [CrossRef]
14. Ehrlich, P.; Hanski, I. *On the Wings of Checkerspots: A Model System for Population Biology*; Oxford University Press: Oxford, UK, 2004. [CrossRef]
15. Schowalter, T. *Insect Ecology—An Ecosystem Approach*, 4th ed.; Academic Press: Cambridge, MA, USA, 2016. [CrossRef]
16. Murdoch, W.; Oaten, A. Predation and Population Stability. In *Advances in Ecological Research*; Academic Press: Cambridge, MA, USA, 1975; Volume 9, pp. 1–131. [CrossRef]

Article

Inverse Problem for the Sobolev Type Equation of Higher Order

Alyona Zamyshlyaeva * and Aleksandr Lut

Department of Applied Mathematics and Programming, South Ural State University,
454080 Chelyabinsk, Russia; lutav@susu.ru
* Correspondence: zamyshliaevaaa@susu.ru

Abstract: The article investigates the inverse problem for a complete, inhomogeneous, higher-order Sobolev type equation, together with the Cauchy and overdetermination conditions. This problem was reduced to two equivalent problems in the aggregate: regular and singular. For these problems, the theory of polynomially bounded operator pencils is used. The unknown coefficient of the original equation is restored using the method of successive approximations. The main result of this work is a theorem on the unique solvability of the original problem. This study continues and generalizes the authors' previous research in this area. All the obtained results can be applied to the mathematical modeling of various processes and phenomena that fit the problem under study.

Keywords: Sobolev type equation; inverse problem; high-order equation; method of successive approximations; polynomial boundedness of operator pencils

1. Introduction

Let $\mathcal{U}, \mathcal{F}, \mathcal{Y}$ be Banach spaces, operators $A, B_0, B_1, ..., B_{n-1} \in \mathcal{L}(\mathcal{U}; \mathcal{F})$, i.e., linear and continuous operators defined on $\mathcal{U}$ and acting to $\mathcal{F}$, ker $A \neq \{0\}$, $C \in \mathcal{L}(\mathcal{U}; \mathcal{Y})$, given functions $\chi : [0, T] \to \mathcal{L}(\mathcal{Y}; \mathcal{F})$, $f : [0, T] \to \mathcal{F}$, $\Psi : [0, T] \to \mathcal{Y}$. Consider the following problem with $t \in [0, T]$

$$Av^{(n)}(t) = B_{n-1} v^{(n-1)}(t) + ... + B_1 v'(t) + B_0 v(t) + q(t)\chi(t) + f(t), \quad (1)$$

$$v(0) = v_0, \ v'(0) = v_1, \ ..., \ v^{(n-1)}(0) = v_{n-1}, \quad (2)$$

$$Cv(t) = \Psi(t). \quad (3)$$

The problem of finding a pair of functions $v(t) \in C^n([0, T]; \mathcal{U})$ (n times continuously diferentiable) and $q(t) \in C^1([0, T]; \mathcal{Y})$ (continuously diferentiable) from relations (1)–(3) is called an inverse problem. At present, the authors have obtained the result when studying the inverse problem, but only in the case of the second-order, Sobolev type equation [1].

The degeneracy of the operator A allows us to classify Equation (1) as a Sobolev type equation. Additionally, one can see that this equation is complete, since all the components $v(t), v'(t), ..., v^{(n)}(t)$ are present. In addition, the Cauchy condition (2) is posed. The overdetermination condition (3) arises due to the need to restore the parameter $q(t)$ of the equation.

The study of Sobolev type equations was carried out repeatedly [1–12]. There are articles devoted to both the first [2–4], the second [1,5,6], the third [7], and the higher [8–10] order. In [2], sufficient conditions for the existence of positive solutions to the Showalter–Sidorov and the Cauchy problem for an abstract linear equation of this type were presented. The linear representatives of Sobolev type equations, such as the Barenblatt–Zheltov–Kochina equation and the Hoff equation are studied in [3]. The paper [7] contains a condition for the existence of a weak, local, timely solution to the Cauchy problem for a model Sobolev type equation. In the study of the direct problem for a higher-order, Sobolev type equation, the phase space method was used [10]. Papers [11,12] are among the first

investigations of Sobolev type equations, and the recent works devoted to applications of Sobolev type equations to real-life models are as follows: [13,14].

The works [1,5,15–24] were devoted to the consideration of inverse problems. In [15], the process of unsteady flow of a viscous incompressible fluid in a pipe with a permeable wall was considered. The dependence on the choice of the boundary of the rectangular region and the unique solvability of the inverse problem were investigated in [16]. The uniqueness criterion for the Lavrent'ev–Bitsadze equation is established in [17]. The correctness in Sobolev spaces of the problem of determining the function of sources in the heat and mass transfer Navier–Stokes system was proved [18]. The problem was finding the area where the vector of boundary displacements and forces is given in parametric form [19]. In [20], the inverse boundary value problem for the heat equation was studied, and the error of the obtained approximate solution was estimated.

The article consists of four sections. The second section combines the necessary, previously obtained, results of the theory of polynomially A-bounded of operator pencils formulated in the form of definitions, theorems and lemmas. Section «Results» has three subsections. The first one presents the result of applying the splitting theorem; thus, the original problem is divided into two equivalent problems in the aggregate: regular and singular. In the second subsection, we study the unique solvability of the regular problem by reducing it to an equivalent problem of the first order and achieving the necessary smoothness for the required function q using the method of successive approximations. The third subsection generalizes the result of studying the singular problem obtained earlier in the work [9], thus obtaining the theorem on the existence and uniqueness of the solution to the problem (1)–(3). In the last section, the significance of the obtained results is given in both the development of the studied theory and their practical application.

2. Preliminary Information

To find a pair of functions $v(t)$ and $q(t)$, we use the results obtained in the research into higher-order, Sobolev type equations [8]. Thus, we will apply the theory of polynomially A-bounded operator pencils. Denote by $\vec{B}$ the pencil of operators $B_0, B_1, ..., B_{n-1}$.

Definition 1. *The sets*

$$\rho^A(\vec{B}) = \{\mu \in \mathbb{C} : (\mu^n A - \mu^{n-1} B_{n-1} - ... - \mu B_1 - B_0)^{-1} \in \mathcal{L}(\mathcal{F};\mathcal{U})\}$$

and $\sigma^A(\vec{B}) = \bar{\mathbb{C}} \backslash \rho^A(\vec{B})$ will be called the A-resolvent set and the A-spectrum of the pencil $\vec{B}$, respectively.

Definition 2. *The operator-function complex variable*

$$R^A_\mu(\vec{B}) = (\mu^n A - \mu^{n-1} B_{n-1} - ... - \mu B_1 - B_0)^{-1}$$

with domain $\rho^A(\vec{B})$ will be called the A-resolvent of the pencil $\vec{B}$.

Definition 3. *Let the pencil $\vec{B}$ be polynomially A-bounded if*

$$\exists a \in \mathbb{R}_+ \ \forall \mu \in \mathbb{C} \ (|\mu| > a) \Rightarrow \left(R^A_\mu(\vec{B}) \in \mathcal{L}(\mathcal{F};\mathcal{U})\right).$$

Let the pencil $\vec{B}$ be polynomially A-bounded. Introduce an important condition

$$\int_\gamma \mu^k R^A_\mu(\vec{B}) d\mu \equiv \mathbb{O}, \ k = 0, 1, ..., n-2, \quad (4)$$

where $\gamma = \{\mu \in \mathbb{C} : |\mu| = r > a\}$.

Lemma 1. *Let the pencil $\vec{B}$ be polynomially A-bounded and condition (4) be fulfilled. Then, the operators*

$$P = \frac{1}{2\pi i}\int_\gamma R_\mu^A(\vec{B})\mu^{n-1}Ad\mu \in \mathcal{L}(\mathcal{U}),$$

$$Q = \frac{1}{2\pi i}\int_\gamma \mu^{n-1}AR_\mu^A(\vec{B})d\mu \in \mathcal{L}(\mathcal{F})$$

are projectors.

Put $\mathcal{U}^0 = \ker P$, $\mathcal{F}^0 = \ker Q$, $\mathcal{U}^1 = \text{im } P$, $\mathcal{F}^1 = \text{im } Q$. From the previous Lemma it follows that $\mathcal{U} = \mathcal{U}^0 \oplus \mathcal{U}^1$, $\mathcal{F} = \mathcal{F}^0 \oplus \mathcal{F}^1$. Let $A^k(B_l^k)$ denote the restriction of the operator $A(B_l)$ onto $\mathcal{U}^k$, $k = 0, 1$; $l = 0, 1, ..., n-1$.

Theorem 1. *Let the pencil $\vec{B}$ be polynomially A-bounded and condition (4) be fulfilled. Then, the actions of the operators split:*
1. *$A^k \in \mathcal{L}(\mathcal{U}^k; \mathcal{F}^k)$, $k = 0, 1$;*
2. *$B_l^k \in \mathcal{L}(\mathcal{U}^k; \mathcal{F}^k)$, $k = 0, 1$; $l = 0, 1, ..., n-1$;*
3. *There exists an operator $(A^1)^{-1} \in \mathcal{L}(\mathcal{F}^1; \mathcal{U}^1)$;*
4. *There exists an operator $(B_0^0)^{-1} \in \mathcal{L}(\mathcal{F}^0; \mathcal{U}^0)$.*

Definition 4. *Define the family of operators $\{K_q^1, K_q^2, ..., K_q^n\}$ as follows*

$$K_1^1 = H_0, \ K_1^2 = -H_1, \ ..., \ K_1^n = -H_{n-1},$$

$$K_{q+1}^1 = K_q^n H_0, \ K_{q+1}^2 = K_q^1 - K_q^n H_1, \ ..., \ K_{q+1}^n = K_q^{n-1} - K_q^n H_{n-1}; \ q = 1, 2, ...,$$

where $H_0 = (B_0^0)^{-1}A^0$, $H_1 = (B_0^0)^{-1}B_1^0$, ..., $H_{n-1} = (B_0^0)^{-1}B_{n-1}^0$.

Definition 5. *The point ∞ is called*
1. *Removable singular point of the A-resolvent of pencil $\vec{B}$, if $K_1^1 \equiv \mathbb{O}$, $K_1^2 \equiv \mathbb{O}$, ..., $K_1^n \equiv \mathbb{O}$;*
2. *A pole of order $p \in \mathbb{N}$ of the A-resolvent of pencil $\vec{B}$, if $\exists \ p$ such, that $K_p^1 \not\equiv \mathbb{O}$, $K_p^2 \not\equiv \mathbb{O}$, ..., $K_p^n \not\equiv \mathbb{O}$, but $K_{p+1}^1 \equiv \mathbb{O}$, $K_{p+1}^2 \equiv \mathbb{O}$, ..., $K_{p+1}^n \equiv \mathbb{O}$;*
3. *An essentially singular point of the A-resolvent of the pencil $\vec{B}$, if $K_p^n \not\equiv \mathbb{O}$ for any $p \in \mathbb{N}$.*

3. Results

3.1. Reduction of the Initial Inverse Problem

Let the pencil $\vec{B}$ be polynomially A-bounded and condition (4) be fulfilled, then $v(t)$ can be represented as $v(t) = Pv(t) + (I - P)v(t) = u(t) + w(t)$. Suppose that $\mathcal{U}^0 \subset \ker C$. Then, by virtue of Theorem 1 and Lemma 1 problem (1)–(3) is equivalent to the problem of finding the functions $u \in C^n([0, T]; \mathcal{U}^1)$, $\omega \in C^n([0, T]; \mathcal{U}^0)$, $q \in C^1([0, T]; \mathcal{Y})$ from the relations

$$u^{(n)}(t) = S_{n-1}u^{(n-1)}(t) + ... + S_1 u'(t) + S_0 u(t) + q(t)(A^1)^{-1}Q\chi(t) + (A^1)^{-1}Qf(t), \quad (5)$$

$$u(0) = u_0, \ u'(0) = u_1, \ ..., \ u^{(n-1)}(0) = u_{n-1}, \quad (6)$$

$$Cu(t) = \Psi(t) \equiv Cv(t), \quad (7)$$

$$H_0 \omega^{(n)}(t) = H_{n-1}\omega^{(n-1)}(t) + ... + H_2 \omega''(t) + H_1 \omega'(t) + \omega(t) + \\ + q(t)(B_0^0)^{-1}(I - Q)\chi(t) + (B_0^0)^{-1}(I - Q)f(t), \quad (8)$$

$$\omega(0) = \omega_0, \ \omega'(0) = \omega_1, \ ..., \ \omega^{(n-1)}(0) = \omega_{n-1}, \quad (9)$$

where
$$S_0 = (A^1)^{-1}B_0^1, \quad S_1 = (A^1)^{-1}B_1^1, \quad \ldots, \quad S_{n-1} = (A^1)^{-1}B_{n-1}^1,$$
$$u_0 = Pv_0, \quad u_1 = Pv_1, \quad \ldots, \quad u_{n-1} = Pv_{n-1},$$
$$w_0 = (I-P)v_0, \quad w_1 = (I-P)v_1, \quad \ldots, \quad w_{n-1} = (I-P)v_{n-1}, \quad t \in [0,T].$$

The inverse problem (5)–(7) is called regular, and problem (8), (9) is called singular.

3.2. Solution of the Regular Inverse Problem

Rewrite problem (5)–(7) in the notation [25]. Let $\mathcal{X} = \mathcal{U}^1$, operators $S_0, S_1, \ldots, S_{n-1} \in Cl(\mathcal{X})$, $C \in \mathcal{L}(\mathcal{X}, \mathcal{Y})$, operator-function $\Phi : [0,T] \to \mathcal{L}(\mathcal{Y}; \mathcal{X})$, functions $h : [0,T] \to \mathcal{X}, \Psi : [0,T] \to \mathcal{Y}$

$$u^{(n)}(t) = S_{n-1}u^{(n-1)}(t) + \ldots + S_1 u'(t) + S_0 u(t) + q(t)\Phi(t) + h(t), \quad t \in [0,T], \quad (10)$$

$$u(0) = u_0, \quad u'(0) = u_1, \quad \ldots, \quad u^{(n-1)}(0) = u_{n-1}, \quad (11)$$

$$Cu(t) = \Psi(t). \quad (12)$$

Theorem 2. *Let the pencil $\vec{B}$ be polynomially A-bounded and condition (4) be fulfilled; moreover, $C \in \mathcal{L}(\mathcal{X}; \mathcal{Y})$, $\Phi \in C^1([0,T]; \mathcal{L}(\mathcal{Y}; \mathcal{X}))$, $h \in C^1([0,T]; \mathcal{X})$, $\Psi \in C^{n+1}([0,T]; \mathcal{Y})$, for any $t \in [0,T]$ the operator $C\Phi(t)$ be invertible and $(C\Phi)^{-1} \in C^1([0,T]; \mathcal{L}(\mathcal{Y}))$. If the compatibility condition $Cu_{n-1} = \Psi^{n-1}(0)$ is satisfied, then the solution to the inverse problem (10)–(12) exists and is unique in the class of functions $q \in C^1([0,T]; \mathcal{Y})$, $u \in C^n([0,T]; \mathcal{X})$.*

Proof of Theorem 2. Reduce problem (10)–(12) to the problem for the first-order equation

$$z'(t) = Az(t) + q(t)Q(t) + H(t), \quad t \in [0,T], \quad (13)$$

$$z(0) = z_0, \quad (14)$$

$$Bz(t) = \Psi(t), \quad (15)$$

where $z(t) = \begin{pmatrix} u(t) \\ \vdots \\ u^{(n-2)}(t) \\ u^{(n-1)}(t) \end{pmatrix}$, $A = \begin{pmatrix} 0 & I & \cdots & 0 \\ \vdots & \vdots & \ddots & \vdots \\ 0 & 0 & \cdots & I \\ S_0 & S_1 & \cdots & S_{n-1} \end{pmatrix}$, $Q(t) = \begin{pmatrix} 0 \\ \vdots \\ 0 \\ \Phi(t) \end{pmatrix}$,

$H(t) = \begin{pmatrix} 0 \\ \vdots \\ 0 \\ h(t) \end{pmatrix}$, $z(0) = \begin{pmatrix} u(0) \\ \vdots \\ u^{(n-2)}(0) \\ u^{(n-1)}(0) \end{pmatrix}$, $z_0 = \begin{pmatrix} u_0 \\ \vdots \\ u_{n-2} \\ u_{n-1} \end{pmatrix}$, $B = \begin{pmatrix} 0 & \cdots & 0 & C \end{pmatrix}$,

$\Psi(t) = \begin{pmatrix} 0 \\ \vdots \\ 0 \\ \Psi^{(n-1)}(t) \end{pmatrix}$.

Put $R(t) = -(C\Phi(t))^{-1}$. Therefore, all the conditions of Theorem 6.2.3 from [25], are fulfilled, and the function $q(t)$ satisfies the integral equation

$$q(t) = q_0(t) + R(t)\left(CS_0 \int_0^t V_{1,n}(t-s)q(s)\Phi(s)ds + \right.$$
$$\left. +CS_1 \int_0^t V_{2,n}(t-s)q(s)\Phi(s)ds + \ldots + CS_{n-1} \int_0^t V_{n,n}(t-s)q(s)\Phi(s)ds \right), \quad (16)$$

where

$$q_0(t) = -R(t)\Big(\Psi^{(n)}(t) - CS_0V_{1,1}(t)u_0 - CS_1V_{2,1}(t)u_0 - \ldots - CS_{n-1}V_{n,1}(t)u_0 -$$

$$-CS_0V_{1,2}(t)u_1 - CS_1V_{2,2}(t)u_1 - \ldots - CS_{n-1}V_{n,2}(t)u_1 - \ldots -$$

$$-CS_0V_{1,n}(t)u_{n-1} - CS_1V_{2,n}(t)u_{n-1} - \ldots - CS_{n-1}V_{n,n}(t)u_{n-1} -$$

$$-CS_0\int_0^t V_{1,n}(t-s)h(s)ds - CS_1\int_0^t V_{2,n}(t-s)h(s)ds -$$

$$-\ldots - CS_{n-1}\int_0^t V_{n,n}(t-s)h(s)ds - Ch(t)\Big).$$

Thus, there exists a unique solution $q \in C^1([0,T];\mathcal{Y})$, $z \in C^1([0,T];\mathcal{X}^n)$ to the inverse problem (13)–(15). And we obtain that the solution to the regular inverse problem (10)–(12) exists and is unique, with $q \in C^1([0,T];\mathcal{Y})$, $u \in C^n([0,T];\mathcal{X})$. □

In order to obtain a solution to a singular problem, we need a greater smoothness of the function q from the solution of a regular problem than class $C^1([0,T];\mathcal{Y})$. Next, we need the following Lemma from [1].

Lemma 2. *Let* $l \in \mathbb{N}$, $V \in C^{l-1}([0,T];\mathcal{L}(\mathcal{X}))$, $g \in C^l([0,T];\mathcal{X})$. *Then*

$$\left(\int_0^t V(t-s)g(s)ds\right)^{(l)} = \sum_{k=0}^{l-1} V^{(l-k-1)}(t)g^{(k)}(0) + \int_0^t V(t-s)g^{(l)}(s)ds.$$

The following theorem provides sufficient conditions for the existence of a more smooth (as $p \in \mathbb{N}$) solution $q \in C^{p+n}([0,T],\mathcal{Y})$ of a regular problem.

Theorem 3. *Let the pencil $\vec{B}$ be polynomially A-bounded and condition (4) be fulfilled, $p \in \mathbb{N}_0$; moreover, $C \in \mathcal{L}(\mathcal{X};\mathcal{Y})$, $\Phi \in C^{p+n}([0,T];\mathcal{L}(\mathcal{Y};\mathcal{X}))$, $h \in C^{p+n}([0,T];\mathcal{X})$, $\Psi \in C^{p+2n}([0,T];\mathcal{Y})$, for any $t \in [0,T]$ operator $C\Phi(t)$ be invertible, with $(C\Phi)^{-1} \in C^{p+n}([0,T];\mathcal{L}(\mathcal{Y}))$ and the compatibility condition $Cu_{n-1} = \Psi^{(n-1)}(0)$ be satisfied for some $u_{n-1} \in \mathcal{U}^1$. Then there exists and a unique solution of (10)–(12) and $q \in C^{p+n}([0,T];\mathcal{Y})$.*

Proof of Theorem 3. Write the propagators of the homogeneous Equation (10) in a matrix, denoting the resolving group of homogeneous Equation (13)

$$V(t) = \begin{pmatrix} V_{1,1}(t) & V_{1,2}(t) & \cdots & V_{1,n-1}(t) & V_{1,n}(t) \\ V_{2,1}(t) & V_{2,2}(t) & \cdots & V_{2,n-1}(t) & V_{2,n}(t) \\ \vdots & \vdots & \ddots & \vdots & \vdots \\ V_{n-1,1}(t) & V_{n-1,2}(t) & \cdots & V_{n-1,n-1}(t) & V_{n-1,n}(t) \\ V_{n,1}(t) & V_{n,2}(t) & \cdots & V_{n,n-1}(t) & V_{n,n}(t) \end{pmatrix} = \frac{1}{2\pi i}\int_\gamma R^A_\mu(\vec{B}) \times$$

$$\times \begin{pmatrix} \mu^{n-1}A - \mu^{n-2}B_{n-1} - \ldots - B_1 & \mu^{n-2}A - \mu^{n-3}B_{n-1} - \ldots - B_2 & \cdots \\ B_0 & \mu^{n-1}A - \mu^{n-2}B_{n-1} - \ldots - \mu B_2 & \cdots \\ \vdots & \vdots & \ddots \\ \mu^{n-3}B_0 & \mu^{n-3}B_1 + \mu^{n-4}B_0 & \cdots \\ \mu^{n-2}B_0 & \mu^{n-2}B_1 + \mu^{n-3}B_0 & \cdots \end{pmatrix}$$

$$\left(\begin{array}{ccc} \cdots & \mu A - B_{n-1} & \mathbb{I} \\ \cdots & \mu^2 A - \mu B_{n-1} & \mu \mathbb{I} \\ \ddots & \vdots & \vdots \\ \cdots & \mu^{n-1} A - \mu^{n-2} B_{n-1} & \mu^{n-2} \mathbb{I} \\ \cdots & \mu^{n-2} B_{n-2} + \mu^{n-3} B_{n-3} + \ldots + B_0 & \mu^{n-1} \mathbb{I} \end{array}\right) e^{\mu t} d\mu,$$

where $\mathbb{I}$ is the identity operator. Earlier, in the proof of Theorem 2, it was established that the function $q(t)$ satisfies the integral Equation (16). Take the natural number $l \leq p + n$. Assuming that $q \in C^l([0, T]; \mathcal{Y})$ by Lemma 2, we obtain the equality

$$q^{(l)}(t) = q_0^{(l)}(t) + \sum_{k=0}^{l-1} C_l^k R^{(k)}(t) CS_0 \sum_{m=0}^{l-k-1} V_{1,n}^{(l-k-m-1)}(t)(q\Phi)^{(m)}(0) +$$

$$+ \sum_{k=0}^{l} \sum_{m=0}^{l-k} C_l^{k,m} R^{(k)}(t) CS_0 \int_0^t V_{1,n}(t-s) q^{(l-k-m)}(s) \Phi^{(m)}(s) ds +$$

$$+ \sum_{k=0}^{l-1} C_l^k R^{(k)}(t) CS_1 \sum_{m=0}^{l-k-1} V_{2,n}^{(l-k-m-1)}(t)(q\Phi)^{(m)}(0) +$$

$$+ \sum_{k=0}^{l} \sum_{m=0}^{l-k} C_l^{k,m} R^{(k)}(t) CS_1 \int_0^t V_{2,n}(t-s) q^{(l-k-m)}(s) \Phi^{(m)}(s) ds + \ldots +$$

$$+ \sum_{k=0}^{l-1} C_l^k R^{(k)}(t) CS_{n-1} \sum_{m=0}^{l-k-1} V_{n,n}^{(l-k-m-1)}(t)(q\Phi)^{(m)}(0) +$$

$$+ \sum_{k=0}^{l} \sum_{m=0}^{l-k} C_l^{k,m} R^{(k)}(t) CS_{n-1} \int_0^t V_{n,n}(t-s) q^{(l-k-m)}(s) \Phi^{(m)}(s) ds,$$

where $C_l^k = \dfrac{l!}{k!(l-k)!}$, $C_l^{k,m} = \dfrac{l!}{k!m!(l-k-m)!}$ and

$$q_0^{(l)}(t) = -\sum_{k=0}^{l} C_l^k R^{(k)}(t) \bigg(\Psi^{(l-k+n)}(t) -$$

$$- CS_0 V_{1,1}^{(l-k)}(t) u_0 - CS_1 V_{2,1}^{(l-k)}(t) u_0 - \ldots - CS_{n-1} V_{n,1}^{(l-k)}(t) u_0 -$$

$$- CS_0 V_{1,2}^{(l-k)}(t) u_1 - CS_1 V_{2,2}^{(l-k)}(t) u_1 - \ldots - CS_{n-1} V_{n,2}^{(l-k)}(t) u_1 - \ldots -$$

$$- CS_0 V_{1,n}^{(l-k)}(t) u_{n-1} - CS_1 V_{2,n}^{(l-k)}(t) u_{n-1} - \ldots - CS_{n-1} V_{n,n}^{(l-k)}(t) u_{n-1} -$$

$$- CS_0 \int_0^t V_{1,n}(t-s) h^{(l-k)}(s) ds - CS_1 \int_0^t V_{2,n}(t-s) h^{(l-k)}(s) ds - \ldots -$$

$$- CS_{n-1} \int_0^t V_{n,n}(t-s) h^{(l-k)}(s) ds - Ch^{(l-k)}(t) \bigg) +$$

$$+ \sum_{k=0}^{l-1} C_l^k R^{(k)}(t) CS_0 \sum_{m=0}^{l-k-1} V_{1,n}^{(l-k-m-1)}(t) h^{(m)}(0) +$$

$$+ \sum_{k=0}^{l-1} C_l^k R^{(k)}(t) CS_1 \sum_{m=0}^{l-k-1} V_{2,n}^{(l-k-m-1)}(t) h^{(m)}(0) + \ldots +$$

$$+ \sum_{k=0}^{l-1} C_l^k R^{(k)}(t) CS_{n-1} \sum_{m=0}^{l-k-1} V_{n,n}^{(l-k-m-1)}(t) h^{(m)}(0)$$

exists from the conditions of this theorem for $l = 0, 1, ..., p + n$.

Show that $q \in C^{p+n}([0,T], \mathcal{Y})$; for this purpose, denote $r_0 = q_0(0)$, and for $l = 1, 2, ..., p + n$, determine the following values

$$r_l = q_0^{(l)}(0) + \sum_{k=0}^{l-1} C_l^k R^{(k)}(0) CS_0 \sum_{m=0}^{l-k-1} V_{1,n}^{(l-k-m-1)}(0) \sum_{j=0}^{m} C_m^j r_{m-j} \Phi^{(j)}(0) +$$

$$+ \sum_{k=0}^{l-1} C_l^k R^{(k)}(0) CS_1 \sum_{m=0}^{l-k-1} V_{2,n}^{(l-k-m-1)}(0) \sum_{j=0}^{m} C_m^j r_{m-j} \Phi^{(j)}(0) + ... +$$

$$+ \sum_{k=0}^{l-1} C_l^k R^{(k)}(0) CS_{n-1} \sum_{m=0}^{l-k-1} V_{n,n}^{(l-k-m-1)}(0) \sum_{j=0}^{m} C_m^j r_{m-j} \Phi^{(j)}(0).$$

Consider the system of integral equations

$$\tilde{q}_0(t) = q_0(t) + R(t) \left(CS_0 \int_0^t V_{1,n}(t-s) \tilde{q}_0(s) \Phi(s) ds + \right.$$

$$+ CS_1 \int_0^t V_{2,n}(t-s) \tilde{q}_0(s) \Phi(s) ds + ... + CS_{n-1} \int_0^t V_{n,n}(t-s) \tilde{q}_0(s) \Phi(s) ds \right),$$

$$\tilde{q}_l(t) = q_0^{(l)}(t) + \sum_{k=0}^{l-1} C_l^k R^{(k)}(t) CS_0 \sum_{m=0}^{l-k-1} V_{1,n}^{(l-k-m-1)}(t) \sum_{j=0}^{m} C_m^j r_{m-j} \Phi^{(j)}(0) +$$

$$+ \sum_{k=0}^{l-1} C_l^k R^{(k)}(t) CS_1 \sum_{m=0}^{l-k-1} V_{2,n}^{(l-k-m-1)}(t) \sum_{j=0}^{m} C_m^j r_{m-j} \Phi^{(j)}(0) + ... +$$

$$+ \sum_{k=0}^{l-1} C_l^k R^{(k)}(t) CS_{n-1} \sum_{m=0}^{l-k-1} V_{n,n}^{(l-k-m-1)}(t) \sum_{j=0}^{m} C_m^j r_{m-j} \Phi^{(j)}(0) +$$

$$+ \sum_{k=0}^{l} \sum_{m=0}^{l-k} C_l^{k,m} R^{(k)}(t) CS_0 \int_0^t V_{1,n}(t-s) \tilde{q}_{l-k-m}(s) \Phi^{(m)}(s) ds +$$

$$+ \sum_{k=0}^{l} \sum_{m=0}^{l-k} C_l^{k,m} R^{(k)}(t) CS_1 \int_0^t V_{2,n}(t-s) \tilde{q}_{l-k-m}(s) \Phi^{(m)}(s) ds + ... +$$

$$+ \sum_{k=0}^{l} \sum_{m=0}^{l-k} C_l^{k,m} R^{(k)}(t) CS_{n-1} \int_0^t V_{n,n}(t-s) \tilde{q}_{l-k-m}(s) \Phi^{(m)}(s) ds,$$

$$l = 1, 2, ..., p + n. \qquad (17)$$

Reduce (17) to the Volterra equation of the second kind

$$g(t) = g_0(t) + \int_0^t K(t,s) g(s) ds$$

on the space $(C([0,T];\mathcal{Y}))^{p+n+1}$ with a matrix operator function $K(t,s)$, given on the triangle $\Delta = \{(t,s) \in \mathbb{R}^2 : 0 \le t \le T, 0 \le s \le t\}$. By virtue of the continuity of all data of system (17), this has a unique solution

$$(\tilde{q}_0, \tilde{q}_1, ..., \tilde{q}_{p+n}) \in (C([0,T];\mathcal{Y}))^{p+n+1}.$$

This solution will be the limit of the sequence of approximations

$$\tilde{q}_{0,i}(t) = q_0(t) + R(t)\left(CS_0 \int_0^t V_{1,n}(t-s)\tilde{q}_{0,i-1}(s)\Phi(s)ds + \right.$$

$$\left. +CS_1 \int_0^t V_{2,n}(t-s)\tilde{q}_{0,i-1}(s)\Phi(s)ds + ... + CS_{n-1} \int_0^t V_{n,n}(t-s)\tilde{q}_{0,i-1}(s)\Phi(s)ds\right),$$

$$\tilde{q}_{l,i}(t) = q_0^{(l)}(t) + \sum_{k=0}^{l-1} C_l^k R^{(k)}(t)CS_0 \sum_{m=0}^{l-k-1} V_{1,n}^{(l-k-m-1)}(t) \sum_{j=0}^m C_m^j r_{m-j}\Phi^{(j)}(0) +$$

$$+ \sum_{k=0}^{l-1} C_l^k R^{(k)}(t)CS_1 \sum_{m=0}^{l-k-1} V_{2,n}^{(l-k-m-1)}(t) \sum_{j=0}^m C_m^j r_{m-j}\Phi^{(j)}(0) + ... +$$

$$+ \sum_{k=0}^{l-1} C_l^k R^{(k)}(t)CS_{n-1} \sum_{m=0}^{l-k-1} V_{n,n}^{(l-k-m-1)}(t) \sum_{j=0}^m C_m^j r_{m-j}\Phi^{(j)}(0) +$$

$$+ \sum_{k=0}^l \sum_{m=0}^{l-k} C_l^{k,m} R^{(k)}(t)CS_0 \int_0^t V_{1,n}(t-s)\tilde{q}_{l-k-m,i-1}(s)\Phi^{(m)}(s)ds +$$

$$+ \sum_{k=0}^l \sum_{m=0}^{l-k} C_l^{k,m} R^{(k)}(t)CS_1 \int_0^t V_{2,n}(t-s)\tilde{q}_{l-k-m,i-1}(s)\Phi^{(m)}(s)ds + ... +$$

$$+ \sum_{k=0}^l \sum_{m=0}^{l-k} C_l^{k,m} R^{(k)}(t)CS_{n-1} \int_0^t V_{n,n}(t-s)\tilde{q}_{l-k-m,i-1}(s)\Phi^{(m)}(s)ds,$$

$$l = 1, 2, ..., p+n;\ i \in \mathbb{N}, \qquad (18)$$

which for $i \to \infty$ on the interval $[0,T]$ converge uniformly to the functions $\tilde{q}_l$, $l = 0, 1, ..., p+n$. Set the initial approximation $\tilde{q}_{l,0} \equiv 0;\ l = 0, 1, ..., p+n$, then $\tilde{q}_{l+1,0} = \tilde{q}'_{l,0};\ l = 0, 1, ..., p+n-1$. In addition, from (18), it follows that

$$\tilde{q}_{l,i}(0) = r_l;\ l = 0, 1, ..., p+n;\ i \in \mathbb{N}. \qquad (19)$$

Assume that for all $\tau = 1, 2, ..., i$ the equalities $\tilde{q}_{l+1,\tau}(t) = \tilde{q}'_{l,\tau}(t)$, $l = 0, 1, ..., p+n-1$ are true. Then, using Lemma 2 and equalities (18), we obtain

$$\frac{d}{dt}\left(\sum_{k=0}^l \sum_{m=0}^{l-k} C_l^{k,m} R^{(k)}(t)CS_0 \int_0^t V_{1,n}(t-s)\tilde{q}_{l-k-m,i}(s)\Phi^{(m)}(s)ds\right) =$$

$$= \sum_{k=0}^l \sum_{m=0}^{l-k} C_l^{k,m} R^{(k+1)}(t)CS_0 \int_0^t V_{1,n}(t-s)\tilde{q}_{l-k-m,i}(s)\Phi^{(m)}(s)ds +$$

$$+ \sum_{k=0}^l \sum_{m=0}^{l-k} C_l^{k,m} R^{(k)}(t)CS_0 V_{1,n}(t)\tilde{q}_{l-k-m,i}(0)\Phi^{(m)}(0) +$$

$$+ \sum_{k=0}^{l} \sum_{m=0}^{l-k} C_l^{k,m} R^{(k)}(t) CS_0 \int_0^t V_{1,n}(t-s)\tilde{q}_{l-k-m,i}(s)\Phi^{(m+1)}(s)ds+$$

$$+ \sum_{k=0}^{l} \sum_{m=0}^{l-k} C_l^{k,m} R^{(k)}(t) CS_0 \int_0^t V_{1,n}(t-s)\tilde{q}_{l-k-m+1,i}(s)\Phi^{(m)}(s)ds =$$

$$= \sum_{k=1}^{l+1} \sum_{m=0}^{l-k+1} C_l^{k-1,m} R^{(k)}(t) CS_0 \int_0^t V_{1,n}(t-s)\tilde{q}_{l-k-m+1,i}(s)\Phi^{(m)}(s)ds +$$

$$+ \sum_{k=0}^{l} C_l^k R^{(k)}(t) CS_0 V_{1,n}(t) \sum_{m=0}^{l-k} C_{l-k}^m r_{l-k-m} \Phi^{(m)}(0) +$$

$$+ \sum_{k=0}^{l} \sum_{m=1}^{l-k+1} C_l^{k,m-1} R^{(k)}(t) CS_0 \int_0^t V_{1,n}(t-s)\tilde{q}_{l-k-m+1,i}(s)\Phi^{(m)}(s)ds +$$

$$+ \sum_{k=0}^{l} \sum_{m=0}^{l-k} C_l^{k,m} R^{(k)}(t) CS_0 \int_0^t V_{1,n}(t-s)\tilde{q}_{l-k-m+1,i}(s)\Phi^{(m)}(s)ds. \quad (20)$$

Denote by

$$a_{k,m} = R^{(k)}(t) CS_0 \int_0^t V_{1,n}(t-s)\tilde{q}_{l-k-m+1,i}(s)\Phi^{(m)}(s)ds, \ l = 2,3,...,p+n.$$

Taking into account the equalities

$$C_l^k + C_l^{k-1} = C_{l+1}^k, \quad C_l^{k,m} + C_l^{k-1,m} + C_l^{k,m-1} = C_{l+1}^{k,m}$$

we obtain

$$\sum_{k=0}^{l} \sum_{m=0}^{l-k} C_l^{k,m} a_{k,m} + \sum_{k=1}^{l+1} \sum_{m=0}^{l-k+1} C_l^{k-1,m} a_{k,m} + \sum_{k=0}^{l} \sum_{m=1}^{l-k+1} C_l^{k,m-1} a_{k,m} =$$

$$= \left(\sum_{k=1}^{l} \sum_{m=1}^{l-k} C_l^{k,m} a_{k,m} + \sum_{k=1}^{l} C_l^{k,0} a_{k,0} + \sum_{m=0}^{l} C_l^{0,m} a_{0,m} \right) +$$

$$+ \left(\sum_{k=1}^{l} \sum_{m=1}^{l-k} C_l^{k-1,m} a_{k,m} + \sum_{k=1}^{l} C_l^{k-1,0} a_{k,0} + \sum_{k=1}^{l} C_l^{k-1,l-k+1} a_{k,l-k+1} + C_l^{l,0} a_{l+1,0} \right) +$$

$$+ \left(\sum_{k=1}^{l} \sum_{m=1}^{l-k} C_l^{k,m-1} a_{k,m} + \sum_{m=1}^{l+1} C_l^{0,m-1} a_{0,m} + \sum_{k=1}^{l} C_l^{k,l-k} a_{k,l-k+1} \right) =$$

$$= \sum_{k=1}^{l} \sum_{m=1}^{l-k} C_{l+1}^{k,m} a_{k,m} + \sum_{k=1}^{l} C_{l+1}^{k,0} a_{k,0} + \sum_{m=1}^{l} C_{l+1}^{0,m} a_{0,m} + \sum_{k=1}^{l} C_{l+1}^{k,0} a_{k,l-k+1} +$$

$$+ C_l^{0,0} a_{0,0} + C_l^{0,l} a_{0,l+1} + C_l^{l,0} a_{l+1,0} = \sum_{k=0}^{l+1} \sum_{m=0}^{l-k+1} C_{l+1}^{k,m} a_{k,m}. \quad (21)$$

For $l = 0,1$ fullment of (21) can be checked directly.
From (20) and (21), it follows that

$$\frac{d}{dt} \left(\sum_{k=0}^{l} \sum_{m=0}^{l-k} C_l^{k,m} R^{(k)}(t) CS_0 \int_0^t V_{1,n}(t-s)\tilde{q}_{l-k-m,i}(s)\Phi^{(m)}(s)ds \right) =$$

$$= \sum_{k=0}^{l+1} \sum_{m=0}^{l-k+1} C_{l+1}^{k,m} R^{(k)}(t) CS_0 \int_0^t V_{1,n}(t-s) \tilde{q}_{l-k-m+1,i}(s) \Phi^{(m)}(s) ds +$$

$$+ \sum_{k=0}^{l} C_l^k R^{(k)}(t) CS_0 V_{1,n}(t) \sum_{m=0}^{l-k} C_{l-k}^m r_{l-k-m} \Phi^{(m)}(0). \quad (22)$$

Similarly, we obtain the result for the subsequent integral element from (18)

$$\frac{d}{dt} \left(\sum_{k=0}^{l} \sum_{m=0}^{l-k} C_l^{k,m} R^{(k)}(t) CS_1 \int_0^t V_{2,n}(t-s) \tilde{q}_{l-k-m,i}(s) \Phi^{(m)}(s) ds \right) =$$

$$= \sum_{k=0}^{l+1} \sum_{m=0}^{l-k+1} C_{l+1}^{k,m} R^{(k)}(t) CS_1 \int_0^t V_{2,n}(t-s) \tilde{q}_{l-k-m+1,i}(s) \Phi^{(m)}(s) ds +$$

$$+ \sum_{k=0}^{l} C_l^k R^{(k)}(t) CS_1 V_{2,n}(t) \sum_{m=0}^{l-k} C_{l-k}^m r_{l-k-m} \Phi^{(m)}(0). \quad (23)$$

Continuing the procedure for all subsequent integral elements (18), we present the result for the last

$$\frac{d}{dt} \left(\sum_{k=0}^{l} \sum_{m=0}^{l-k} C_l^{k,m} R^{(k)}(t) CS_{n-1} \int_0^t V_{n,n}(t-s) \tilde{q}_{l-k-m,i}(s) \Phi^{(m)}(s) ds \right) =$$

$$= \sum_{k=0}^{l+1} \sum_{m=0}^{l-k+1} C_{l+1}^{k,m} R^{(k)}(t) CS_{n-1} \int_0^t V_{n,n}(t-s) \tilde{q}_{l-k-m+1,i}(s) \Phi^{(m)}(s) ds +$$

$$+ \sum_{k=0}^{l} C_l^k R^{(k)}(t) CS_{n-1} V_{n,n}(t) \sum_{m=0}^{l-k} C_{l-k}^m r_{l-k-m} \Phi^{(m)}(0). \quad (24)$$

Changing the summation indices and re-grading the sums, we obtain

$$\frac{d}{dt} \left(\sum_{k=0}^{l-1} C_l^k R^{(k)}(t) CS_0 \sum_{m=0}^{l-k-1} V_{1,n}^{(l-k-m-1)}(t) \sum_{j=0}^{m} C_m^j r_{m-j} \Phi^{(j)}(0) \right) =$$

$$= \sum_{k=0}^{l-1} C_l^k R^{(k)}(t) CS_0 \sum_{m=0}^{l-k-1} V_{1,n}^{(l-k-m)}(t) \sum_{j=0}^{m} C_m^j r_{m-j} \Phi^{(j)}(0) +$$

$$+ \sum_{k=0}^{l-1} C_l^k R^{(k+1)}(t) CS_0 \sum_{m=0}^{l-k-1} V_{1,n}^{(l-k-m-1)}(t) \sum_{j=0}^{m} C_m^j r_{m-j} \Phi^{(j)}(0) =$$

$$= \sum_{k=0}^{l-1} C_l^k R^{(k)}(t) CS_0 \sum_{m=0}^{l-k-1} V_{1,n}^{(l-k-m)}(t) \sum_{j=0}^{m} C_m^j r_{m-j} \Phi^{(j)}(0) +$$

$$+ \sum_{k=1}^{l} C_l^{k-1} R^{(k)}(t) CS_0 \sum_{m=0}^{l-k} V_{1,n}^{(l-k-m)}(t) \sum_{j=0}^{m} C_m^j r_{m-j} \Phi^{(j)}(0) =$$

$$= \left(\sum_{k=1}^{l-1} C_l^k R^{(k)}(t) CS_0 \sum_{m=0}^{l-k-1} V_{1,n}^{(l-k-m)}(t) \sum_{j=0}^{m} C_m^j r_{m-j} \Phi^{(j)}(0) + \right.$$

$$\left. + C_l^0 R(t) CS_0 \sum_{m=0}^{l-1} V_{1,n}^{(l-m)}(t) \sum_{j=0}^{m} C_m^j r_{m-j} \Phi^{(j)}(0) \right) +$$

$$+\left(\sum_{k=1}^{l-1} C_l^{k-1} R^{(k)}(t) CS_0 \sum_{m=0}^{l-k-1} V_{1,n}^{(l-k-m)}(t) \sum_{j=0}^{m} C_m^j r_{m-j} \Phi^{(j)}(0)+\right.$$

$$+\sum_{k=1}^{l-1} C_l^{k-1} R^{(k)}(t) CS_0 V_{1,n}(t) \sum_{j=0}^{l-k} C_{l-k}^j r_{l-k-j} \Phi^{(j)}(0)+$$

$$\left.+C_l^{l-1} R^{(l)}(t) CS_0 V_{1,n}(t) C_0^0 r_0 \Phi(0)\right) =$$

$$= \sum_{k=0}^{l} C_{l+1}^{k} R^{(k)}(t) CS_0 \sum_{m=0}^{l-k} V_{1,n}^{(l-k-m)}(t) \sum_{j=0}^{m} C_m^j r_{m-j} \Phi^{(j)}(0)-$$

$$- \sum_{k=0}^{l} C_l^{k} R^{(k)}(t) CS_0 V_{1,n}(t) \sum_{m=0}^{l-k} C_{l-k}^m r_{l-k-m} \Phi^{(m)}(0). \qquad (25)$$

Similarly, we obtain the result for the next non-integral element from (18)

$$\frac{d}{dt}\left(\sum_{k=0}^{l-1} C_l^k R^{(k)}(t) CS_1 \sum_{m=0}^{l-k-1} V_{2,n}^{(l-k-m-1)}(t) \sum_{j=0}^{m} C_m^j r_{m-j} \Phi^{(j)}(0)\right) =$$

$$= \sum_{k=0}^{l} C_{l+1}^{k} R^{(k)}(t) CS_1 \sum_{m=0}^{l-k} V_{2,n}^{(l-k-m)}(t) \sum_{j=0}^{m} C_m^j r_{m-j} \Phi^{(j)}(0)-$$

$$- \sum_{k=0}^{l} C_l^{k} R^{(k)}(t) CS_1 V_{2,n}(t) \sum_{m=0}^{l-k} C_{l-k}^m r_{l-k-m} \Phi^{(m)}(0). \qquad (26)$$

Continuing the procedure for all subsequent non-integral elements (18), we present the result for the last

$$\frac{d}{dt}\left(\sum_{k=0}^{l-1} C_l^k R^{(k)}(t) CS_{n-1} \sum_{m=0}^{l-k-1} V_{n,n}^{(l-k-m-1)}(t) \sum_{j=0}^{m} C_m^j r_{m-j} \Phi^{(j)}(0)\right) =$$

$$= \sum_{k=0}^{l} C_{l+1}^{k} R^{(k)}(t) CS_{n-1} \sum_{m=0}^{l-k} V_{n,n}^{(l-k-m)}(t) \sum_{j=0}^{m} C_m^j r_{m-j} \Phi^{(j)}(0)-$$

$$- \sum_{k=0}^{l} C_l^{k} R^{(k)}(t) CS_{n-1} V_{n,n}(t) \sum_{m=0}^{l-k} C_{l-k}^m r_{l-k-m} \Phi^{(m)}(0). \qquad (27)$$

Differentiating (18), and also using (22)–(27), we obtain the equalities $\tilde{q}'_{l,i+1} = \tilde{q}_{l+1,i+1}$; $l = 0, 1, ..., p + n - 1$. Thus, the sequence $\tilde{q}_{0,i}$ converges as $i \to \infty$ to the function $\tilde{q}_0$ uniformly on the interval $[0, T]$, and the sequence $\tilde{q}'_{0,i} = \tilde{q}_{1,i}$ converges as $i \to \infty$ to the function $\tilde{q}_1$ uniformly on the segment $[0, T]$. Therefore, the function $\tilde{q}_0$ is continuously differentiable and $\tilde{q}'_0 = \tilde{q}_1$. The equalities of $\tilde{q}'_l = \tilde{q}_{l+1}$; $l = 1, 2, ..., p + n - 1$, are proved in the same way, which implies that $\tilde{q}_0 \equiv q \in C^{p+n}([0, T]; \mathcal{Y})$ and, therefore, $q^{(l)} = \tilde{q}_l$; $l = 1, 2, ..., p + n$. □

3.3. Solvability of the Original Inverse Problem

Theorem 4. *Let the pencil $\vec{B}$ be polynomially A-bounded and condition (4) be fulfilled; moreover, the ∞ be a pole of order $p \in \mathbb{N}_0$ of the A-resolvent of the pencil $\vec{B}$, operator $C \in \mathcal{L}(\mathcal{U}; \mathcal{Y})$, $\mathcal{U}^0 \subset \ker C$, $\chi \in C^{p+n}([0, T]; \mathcal{L}(\mathcal{Y}; \mathcal{F}))$, $f \in C^{p+n}([0, T]; \mathcal{F})$, $\Psi \in C^{p+2n}([0, T]; \mathcal{Y})$, for any $t \in [0, T]$ operator $C(A^1)^{-1}Q\chi$ be invertible, with $\left(C(A^1)^{-1}Q\chi\right)^{-1} \in C^{p+n}([0, T]; \mathcal{L}(\mathcal{Y}))$, the* condition $Cu_{n-1} = \Psi^{(n-1)}(0)$ be satisfied at some initial value $u_{n-1} \in \mathcal{U}^1$, and the initial values $w_k = (I - P)v_k \in \mathcal{U}^0$ satisfy

$$w_k = -\sum_{j=0}^{p} K_j^n (B_0^0)^{-1} \frac{d^{j+k}}{dt^{j+k}}\left[(I - Q)(q(0)\chi(0) + f(0))\right], \quad k = 0, 1, ..., n-1.$$

Then, there exists a unique solution (v, q) of inverse problem (1)–(3), where $q \in C^{p+n}([0,T]; \mathcal{Y})$, $v = u + w$, whence $u \in C^n([0,T]; \mathcal{U}^1)$ is the solution of (5)–(7), and the function $w \in C^n([0,T]; \mathcal{U}^0)$ is a solution of (8) and (9) given by

$$w(t) = -\sum_{j=0}^{p} K_j^n (B_0^0)^{-1} \frac{d^j}{dt^j}\left[(I - Q)(q(t)\chi(t) + f(t))\right]. \tag{28}$$

Proof of Theorem 4. The conditions of Theorems 2 and 3 are satisfied, and, therefore, there exists a unique solution (q, u) to problem (5)–(7), where $q \in C^{p+n}([0,T]; \mathcal{Y})$, $u \in C^n([0,T]; \mathcal{U}^1)$.

Using the result of [9] and the required smoothness of the function q, we obtain that there exists a unique solution $w \in C^n([0,T]; \mathcal{U}^0)$ to (8), (9), given by (28). □

4. Discussion

The results obtained in the article can be applied to various mathematical models, such as a model of oscillation of a rotating viscous fluid using the viscosity coefficient, a model of gravitational-gyroscopic and internal waves, and a model of sound waves in smectics, since these mathematical models can be reduced to the Sobolev type equations of higher order. One of the most typical examples of the application of the Sobolev type equations theory is the Boussinesq–Love model [5]:

$$(\lambda - \Delta)v_{tt} = \alpha(\Delta - \lambda')v_t + \beta(\Delta - \lambda'')v + qf, \tag{29}$$

with initial conditions

$$v(x,0) = v_0(x), \quad v_t(x,0) = v_1(x),$$

boundary condition

$$v(x,t)|_{\partial \Omega} = 0$$

and overdetermination condition

$$\int_{\Omega} v(x,t) K(x) dx = \Phi(t), \tag{30}$$

where $v_0(x), v_1(x), K(x), \Phi(t)$ are given functions, $v(x, t)$ is a searched function and $\Omega \subset \mathbb{R}^n$ is a bounded domain with a boundary $\partial \Omega$ of class C^∞. Equation (29) describes longitudinal vibrations in a thin elastic rod, taking into account the inertia and external load. The coefficients $\lambda, \alpha, \lambda', \beta, \lambda''$ characterize the properties of the rod material and relate such quantities as Young's modulus, Poisson's ratio, material density and radius of gyration relative to the center of gravity, in addition, the function f sets a known part of the external load (if known). The integral overdetermination condition (30) arises at the moment when, in addition to finding the function v, it is necessary to restore the component of the external load q. In addition, it is planned to use the obtained results for the development of numerical methods, to find approximate solutions to some of the previously presented models.

Author Contributions: Conceptualization, A.Z. and A.L.; Methodology, A.Z.; Validation, A.Z. and A.L.; Formal Analysis, A.L.; Investigation, A.L.; Resources, A.Z.; Data Curation, A.L.; Writing—Original Draft Preparation, A.L.; Supervision, A.Z.; Project administration, A.Z.; Funding Acquisition, A.L. All authors have read and agreed to the published version of the manuscript.

Funding: The reported study was funded by RFBR, project number 19-31-90137.

Institutional Review Board Statement: Not applicable.

Informed Consent Statement: Not applicable.

Data Availability Statement: Not applicable.

Conflicts of Interest: The authors declare no conflict of interest.

References

1. Zamyshlyaeva, A.A.; Lut, A.V. Inverse Problem for Sobolev Type Mathematical Models. *Bull. South Ural State Univ. Ser. Math. Model. Program.* **2019**, *12*, 25–36. [CrossRef]
2. Banasiak, J.; Manakova, N.A.; Sviridyuk, G.A. Positive solutions to Sobolev type equations with relatively p-sectorial operators. *Bull. South Ural. State Univ. Ser. Math. Model. Program.* **2020**, *13*, 17–32. [CrossRef]
3. Shafranov, D.E. On numerical solution in the space of differential forms for one stochastic sobolev-type equation with a relatively radial operator. *J. Comput. Eng. Math.* **2020**, *7*, 48–55. [CrossRef]
4. Shafranov, D.E.; Adukova, N.V. Solvability of the Showalter–Sidorov problem for Sobolev type equations with operators in the form of first-order polynomials from the Laplace–Beltrami operator on differential forms. *J. Comput. Eng. Math.* **2017**, *4*, 27–34. [CrossRef]
5. Lut, A.V. Numerical Investigation of The Inverse Problem for The Boussinesq–Love Mathematical Model. *J. Comput. Eng. Math.* **2020**, *7*, 45–59. [CrossRef]
6. Zamyshlyaeva, A.A.; Lut, A.V. Numerical investigation of the Boussinesq–Love mathematical models on geometrical graphs. *Bull. South Ural State Univ. Ser. Math. Model. Program.* **2017**, *10*, 137–143. [CrossRef]
7. Korpusov, M.O.; Panin, A.A.; Shishkov, A.E. On the Critical Exponent «Instantaneous Blow-up» Versus «Local Solubility» in the Cauchy Problem for a Model Equation of Sobolev Type. *Izv. Math.* **2021**, *85*, 111–144.
8. Zamyshlyaeva, A.A.; Bychkov, E.V. The Cauchy Problem for the Sobolev Type Equation of Higher Order. *Bull. South Ural State Univ. Ser. Math. Model. Program.* **2018**, *11*, 5–14. [CrossRef]
9. Zamyshlyaeva, A.A.; Tsyplenkova, O.N.; Bychkov, E.V. Optimal Control of Solutions to the Initial-final Problem for the Sobolev Type Equation of Higher Order. *J. Comput. Eng. Math.* **2016**, *3*, 57–67. [CrossRef]
10. Zamyshlyaeva, A.A.; Manakova, N.A.; Tsyplenkova, O.N. Optimal Control in Linear Sobolev Type Mathematical Models. *Bull. South Ural State Univ. Ser. Math. Model. Program.* **2020**, *13*, 5–27.
11. Sobolev, S.L. On a new problem of mathematical physics. *Izv. Akad. Nauk SSSR Ser. Mat.* **1954**, *18*, 3–50.
12. Showalter, R.E. The Sobolev type equations. I. *Appl. Anal.* **1975**, *5*, 15–22. [CrossRef]
13. Mohan, M.T. On the three dimensional Kelvin-Voigt fluids: Global solvability, exponential stability and exact controllability of Galerkin approximations. *Evol. Equ. Control Theory* **2020**, *9*, 301–339. [CrossRef]
14. Baranovskii, E.S. Strong solutions of the incompressible Navier–Stokes–Voigt model. *Mathematics* **2020**, *8*, 181. [CrossRef]
15. Gamzaev, K.M. Inverse Problem of Unsteady Incompressible Fluid Flow in a Pipe with a Permeable Wall. *Bull. South Ural State Univ. Ser. Math. Mech. Phys.* **2020**, *12*, 24–30.
16. Islomov, B.I.; Ubaydullayev, U.S. The Inverse Problem for a Mixed Type Equation with a Fractional Order Operator in a Rectangular Domain. *Russ. Math.* **2021**, *65*, 25–42. [CrossRef]
17. Martemyanova, N.V. Nonlocal Inverse Problem of Finding Unknown Multipliers in the Right-Hand Part of Lavrentiev–Bitsadze Equation. *Russ. Math.* **2020**, *64*, 40–57. [CrossRef]
18. Pyatkov, S.G. On Some Classes of Inverse Problems on Determining Source Functions for Heat and Mass Transfer Systems. *J. Math. Sci.* **2020**, *188*, 23–42.
19. Shirokova, E.A. Inverse Boundary Value Problem of the Plane Theory of Elasticity. *Russ. Math.* **2020**, *64*, 66–73. [CrossRef]
20. Tanana, V.P. On Reducing an Inverse Boundary-Value Problem to the Synthesis of Two Ill-Posed Problems and Their Solution. *Numer. Anal. Appl.* **2020**, *13*, 180–192. [CrossRef]
21. Baev, A.V. On the Solution of an Inverse Problem for Equations of Shallow Water in a Pool with Variable Depth. *Math. Model. Comput. Simulations* **2020**, *32*, 3–15.
22. Glasko, Y.V. The Inverse Problem of Interpretation of Gravitational and Magnetic Anomalies of Hydrocarbon Deposits. *J. Appl. Ind. Math.* **2020**, *14*, 46–55. [CrossRef]
23. Shimelevich, M.I. On the Method of Calculating the Modulus of Continuity of The Inverse Operator and its Modifications with Application to Non-linear Problems of Geoelectrics. *Numer. Methods Program.* **2020**, *21*, 350–372.
24. Timonov, A. A novel method for the numerical solution of a hybrid inverse problem of electrical conductivity imaging. *Investig. Appl. Math. Inform. Part I Zap. Nauchnykh Semin. POMI* **2021**, *499*, 105–128.
25. Prilepko, A.I.; Orlovsky, D.G.; Vasin, I.A. *Methods for Solving Inverse Problems in Mathematical Physics*; Marcel Dekker: New York, NY, USA, 2000.

Article
Important Criteria for Asymptotic Properties of Nonlinear Differential Equations

Ahmed AlGhamdi [1,†], Omar Bazighifan [2,3,†] and Rami Ahmad El-Nabulsi [4,5,6,*,†]

1. Department of Computer Engineering, College of Computers and Information Technology, Taif University, Taif 21944, Saudi Arabia; asjannah@tu.edu.sa
2. Section of Mathematics, International Telematic University Uninettuno, CorsoVittorio Emanuele II, 39, 00186 Roma, Italy; o.bazighifan@gmail.com
3. Department of Mathematics, Faculty of Science, Hadhramout University, Hadhramout 50512, Yemen
4. Research Center for Quantum Technology, Faculty of Science, Chiang Mai University, Chiang Mai 50200, Thailand
5. Department of Physics and Materials Science, Faculty of Science, Chiang Mai University, Chiang Mai 50200, Thailand
6. Athens Institute for Education and Research, Mathematics and Physics Divisions, 8 Valaoritou Street, Kolonaki, 10671 Athens, Greece
* Correspondence: el-nabulsi@atiner.gr or nabulsiahmadrami@yahoo.fr
† These authors contributed equally to this work.

Abstract: In this article, we prove some new oscillation theorems for fourth-order differential equations. New oscillation results are established that complement related contributions to the subject. We use the Riccati technique and the integral averaging technique to prove our results. As proof of the effectiveness of the new criteria, we offer more than one practical example.

Keywords: fourth-order differential equations; neutral delay; oscillation

1. Introduction

In this manuscript, we are concerned with the asymptotic behavior of solutions to fourth-order differential equations:

$$\left(m(z)\Psi_{r_1}\left(\varsigma'''(z)\right)\right)' + \tilde{\omega}(z)\Psi_{r_2}(\delta(\alpha(z))) = 0, \tag{1}$$

where $\Psi_{r_i}[s] = |s|^{r_i-1}s, i = 1,2$, $\varsigma(z) = \delta(z) + \tilde{y}(z)\delta(\tilde{\alpha}(z))$, $m, \tilde{y}, \tilde{\omega} \in C[z_0, \infty)$, $m(z) > 0$, $m'(z) \geq 0$, $\tilde{\omega}(z) > 0$, $0 \leq \tilde{y}(z) < \tilde{y}_0 < \infty$, $\tilde{\alpha}, \alpha \in C[z_0, \infty)$, $\tilde{\alpha}(z) \leq z$, $\lim_{z\to\infty} \tilde{\alpha}(z) = \lim_{z\to\infty} \alpha(z) = \infty$; and r_1 and r_2 are quotients of odd positive integers, under the assumption of the following:

$$\int_{z_0}^{\infty} \frac{1}{m^{1/r_1}(s)} ds = \infty. \tag{2}$$

The theory of the oscillation of delay of differential equations is a fertile study area and has attracted the attention of many authors recently. This is due to the existence of many important applications of this theory in neural networks, biology, social sciences, engineering, etc.; see [1,2].

A study of the behavior of solutions to higher order differential equations yields much fewer results than for the least order equations although they are of the utmost importance in a lot of applications, especially neutral delay differential equations.

Currently, there are studies on the oscillation results of differential equations, so many of these studies have been devoted to study the oscillation of different classes of differential equations by using different techniques in order to establish sufficient conditions to ensure the oscillatory behavior of the solutions of (1), see [3–5].

The motivation for studying this article is complemented by the results reported in [6,7]; therefore, we discuss their findings and results below.

Xing et al. [6] presented criteria for oscillation of the equation as follows:

$$\left(m(z)\left(\varsigma^{(n-1)}(z)\right)^{r_1}\right)' + \tilde{\omega}(z)\delta^{r_1}(\alpha(z)) = 0,$$

under the conditions

$$\left(\alpha^{-1}(z)\right)' \geq \alpha_0 > 0, \ \tilde{\alpha}'(z) \geq \tilde{\alpha}_0 > 0, \ \tilde{\alpha}^{-1}(\alpha(z)) < z$$

and

$$\liminf_{z \to \infty} \int_{\tilde{\alpha}^{-1}(\alpha(z))}^{z} \frac{\hat{\omega}(s)}{m(s)} \left(s^{n-1}\right)^{r_1} ds > \left(\frac{1}{\alpha_0} + \frac{\tilde{y}_0^{r_1}}{\alpha_0 \tilde{\alpha}_0}\right) > \frac{((n-1)!)^{r_1}}{e},$$

where $0 \leq \tilde{y}(z) < \tilde{y}_0 < \infty$ and $\hat{\omega}(z) := \min\{\tilde{\omega}(\alpha^{-1}(z)), \tilde{\omega}(\alpha^{-1}(\tilde{\alpha}(z)))\}$. Moreover, the authors used the comparison method to obtain oscillation conditions for this equation.

Bazighifan et al. [7] presented oscillation results for the following fourth-order equation:

$$\left(m(z)\left(\varsigma'''(z)\right)^{r_1}\right)' + \tilde{\omega}(z)\delta^{r_1}(\alpha(z)) = 0,$$

under the conditions

$$\int_{z_0}^{\infty} \frac{1}{m^{1/r_1}(s)} ds < \infty$$

using the Riccati technique.

Zhang et al. [8] established oscillation criteria for the following equation:

$$\left(m(z)\left(\varsigma^{(n-1)}(z)\right)^{r_1}\right)' + \tilde{\omega}(z)f(\delta(\alpha(z)))ds = 0$$

and under the condition

$$\int_{z_0}^{\infty} \left(k\rho(z)E(z) - \frac{1}{4\lambda}\left(\frac{\rho'(z)}{\rho(z)}\right)^2 \eta(z)\right) ds = \infty.$$

Chatzarakis et al. [9], by using the Riccati technique, established asymptotic behavior for the following neutral equation:

$$\left(m(z)\left(\varsigma'''(z)\right)^{r_1}\right)' + \int_a^b \tilde{\omega}(z,s)f(\delta(\alpha(z,s)))ds = 0.$$

The authors in [6,7] used the comparison technique that differs from the one we used in this article. Their approach is based on using these mentioned methods to reduce Equation (1) into a first-order equation, while in our article, we discuss the oscillatory properties of differential equations with a middle term and with a canonical operator of the neutral-type, and we employ a different approach based on using the integral averaging technique and the Riccati technique to reduce the main equation into a first-order inequality to obtain more effective oscillatory properties.

The purpose of this article is to establish new oscillation criteria for (1). The methods used in this paper simplify and extend some of the known results that are reported in the literature [6,7]. The authors in [6,7] used a comparison technique that differs from the one we used in this article.

2. Oscillation Criteria

We next present the lemmas needed for the proof of the original results:

Lemma 1 ([10]). *If $\delta^{(i)}(z) > 0$, $i = 0, 1, \ldots, n$, and $\delta^{(n+1)}(z) < 0$, then the following holds:*

$$n! \frac{\delta(z)}{z^n} \geq (n-1)! \frac{\delta'(z)}{z^{n-1}}.$$

Lemma 2 ([11]). *Let $\delta \in C^n([z_0, \infty), (0, \infty))$. Assume that $\delta^{(n)}(z)$ is of fixed sign and not identically zero on $[z_0, \infty)$ and that there exists a $z_1 \geq z_0$ such that $\delta^{(n-1)}(z)\delta^{(n)}(z) \leq 0$ for all $z \geq z_1$. If $\lim_{z \to \infty} \delta(z) \neq 0$, then for every $\mu \in (0,1)$ there exists $z_\mu \geq z_1$ such that the following holds:*

$$\delta(z) \geq \frac{\mu}{(n-1)!} z^{n-1} \left| \delta^{(n-1)}(z) \right| \text{ for } z \geq z_\mu.$$

Lemma 3 ([12]). *Let $a \geq 0$; then, the following holds:*

$$X\delta - Y\delta^{(a+1)/a} \leq a^a(a+1)^{-(a+1)} Y^{-a} X^{a+1},$$

where $Y > 0$ and X are constants.

Lemma 4 ([13]). *Assume that $\delta(z)$ is an eventually positive solution of Equation (1). Then,*

Case $(\mathbf{N}_1)$: $\varsigma(z) > 0, \varsigma'(z) > 0, \varsigma''(z) > 0, \varsigma'''(z) > 0$,
Case $(\mathbf{N}_2)$: $\varsigma(z) > 0, \varsigma'(z) > 0, \varsigma''(z) < 0, \varsigma'''(z) > 0$.

Here are the notations used for our study:

$$E_1(z) = \beta(z)\tilde{\omega}(z)(1 - \tilde{y}_0)^{r_2} A_1^{r_2 - r_1} \left(\frac{\alpha(z)}{z} \right)^{3r_2},$$

$$\Phi(z) = (1 - \tilde{y}_0)^{r_2/r_1} h(z) A_2^{r_2/r_1 - 1}(z) \int_z^\infty \left(\frac{1}{m(u)} \int_u^\infty \tilde{\omega}(s) \frac{\alpha^{r_2}(s)}{s^{r_2}} ds \right)^{1/r_1} du$$

and

$$\Theta(z) = r_1 \mu_1 \frac{z^2}{2m^{1/r_1}(z)\beta^{1/r_1}(z)}.$$

Lemma 5. *Let $\delta(z)$ is an eventually positive solution of Equation (1), then*

$$\left(m(z) \left(\varsigma'''(z) \right)^{r_1} \right)' \leq -G(z) \left(\varsigma'''(\alpha(z)) \right)^{r_2}, \tag{3}$$

where

$$G(z) = \tilde{\omega}(z)(1 - \tilde{y}_0)^{r_2} \left(\frac{\mu}{6} (\alpha(z))^3 \right)^{r_2}.$$

Proof. Let $\delta(z)$ is an eventually positive solution of Equation (1). From definition of $\varsigma(z) = \delta(z) + \tilde{y}(z)\delta(\tilde{\alpha}(z))$, we obtain the following:

$$\begin{aligned} \delta(z) &\geq \varsigma(z) - \tilde{y}_0 \delta(\tilde{\alpha}(z)) \\ &\geq \varsigma(z) - \tilde{y}_0 \varsigma(\tilde{\alpha}(z)) \\ &\geq (1 - \tilde{y}_0)\varsigma(z), \end{aligned}$$

which with (1), results in the following:

$$\left(m(z)\left(\varsigma'''(z)\right)^{r_1} \right)' + \tilde{\omega}(z)(1 - \tilde{y}_0)^{r_2} \varsigma^{r_2}(\alpha(z)) \leq 0. \tag{4}$$

Using Lemma 2, we see the following:

$$\varsigma(z) \geq \frac{\mu}{6} z^3 \varsigma'''(z). \tag{5}$$

Combining (4) and (5), we find the following:

$$\left(m(z)(\varsigma'''(z))^{r_1}\right)' + \tilde{\omega}(z)(1-\tilde{y}_0)^{r_2}\left(\frac{\mu}{6}(\alpha(z))^3\right)^{r_2}(\varsigma'''(\alpha(z)))^{r_2} \leq 0.$$

Thus, (3) holds. This completes the proof. □

Lemma 6. *Let $\delta(z)$ is an eventually positive solution of Equation (1) and*

$$B'(z) \leq \frac{\beta'(z)}{\beta(z)}B(z) - E_1(z) - r_1\mu_1\frac{z^2}{2m^{1/r_1}(z)\beta^{1/r_1}(z)}B^{\frac{r_1+1}{r_1}}(z), \text{ if } \varsigma \text{ satisfies } (\mathbf{N}_1) \quad (6)$$

and

$$A'(z) \leq -\Phi(z) + \frac{h'(z)}{h(z)}A(z) - \frac{1}{h(z)}A^2(z), \text{ if } \varsigma \text{ satisfies } (\mathbf{N}_2), \quad (7)$$

where

$$B(z) := \beta(z)\frac{m(z)(\varsigma'''(z))^{r_1}}{\varsigma^{r_1}(z)} > 0 \quad (8)$$

and

$$A(z) := h(z)\frac{\varsigma'(z)}{\varsigma(z)}, \; z \geq z_1. \quad (9)$$

Proof. Let $\delta(z)$ is an eventually positive solution of Equation (1). Let $(\mathbf{N}_1)$ holds. From (8) and (4), we find the following:

$$B'(z) \leq \frac{\beta'(z)}{\beta(z)}B(z) - \beta(z)\tilde{\omega}(z)(1-\tilde{y}_0)^{r_2}\frac{\varsigma^{r_2}(\alpha(z))}{\varsigma^{r_1}(z)} - r_1\beta(z)\frac{m(z)(\varsigma'''(z))^{r_1}}{\varsigma^{r_1+1}(z)}\varsigma'(z). \quad (10)$$

Using Lemma 1, we find

$$\varsigma(z) \geq \frac{z}{3}\varsigma'(z)$$

and hence,

$$\frac{\varsigma(\alpha(z))}{\varsigma(z)} \geq \frac{\alpha^3(z)}{z^3}. \quad (11)$$

It follows from Lemma 2 that

$$\varsigma'(z) \geq \frac{\mu_1}{2}z^2\varsigma'''(z), \quad (12)$$

for all $\mu_1 \in (0,1)$ and every sufficiently large z. Thus, by (10)–(12), we obtain the following:

$$B'(z) \leq \frac{\beta'(z)}{\beta(z)}B(z) - \beta(z)\tilde{\omega}(z)(1-\tilde{y}_0)^{r_2}\varsigma^{r_2-r_1}(z)\left(\frac{\alpha(z)}{z}\right)^{3r_2}$$
$$-r_1\mu_1\frac{z^2}{2m^{1/r_1}(z)\beta^{1/r_1}(z)}B^{\frac{r_1+1}{r_1}}(z).$$

Since $\varsigma'(z) > 0$, there exist $z_2 \geq z_1$ and $A_1 > 0$ such that the following holds:

$$\varsigma(z) > A_1. \quad (13)$$

Thus, we obtain the following:

$$B'(z) \leq \frac{\beta'(z)}{\beta(z)}B(z) - \beta(z)\tilde{\omega}(z)(1-\tilde{y}_0)^{r_2}A^{r_2-r_1}\left(\frac{\alpha(z)}{z}\right)^{3r_2}$$
$$-r_1\mu_1\frac{z^2}{2m^{1/r_1}(z)\beta^{1/r_1}(z)}B^{\frac{r_1+1}{r_1}}(z),$$

which yields the following:

$$B'(z) \leq \frac{\beta'(z)}{\beta(z)}B(z) - E_1(z) - r_1\mu_1\frac{z^2}{2m^{1/r_1}(z)\beta^{1/r_1}(z)}B^{\frac{r_1+1}{r_1}}(z).$$

Thus, (6) holds.
Let (N_2) hold. Integrating (4) from z to u, we find the following:

$$m(u)(\varsigma'''(u))^{r_1} - m(z)(\varsigma'''(z))^{r_1} \leq -\int_z^u \tilde{\omega}(s)(1-\tilde{y}_0)^{r_2}\varsigma^{r_2}(\alpha(s))ds. \tag{14}$$

From Lemma 1, we obtain the following:

$$\varsigma(z) \geq z\varsigma'(z)$$

and hence,

$$\varsigma(\alpha(z)) \geq \frac{\alpha(z)}{z}\varsigma(z). \tag{15}$$

For (14), letting $u \to \infty$ and using (15), we obtain the following:

$$m(z)(\varsigma'''(z))^{r_1} \geq (1-\tilde{y}_0)^{r_2}\varsigma^{r_2}(z)\int_z^\infty \tilde{\omega}(s)\frac{\alpha^{r_2}(s)}{s^{r_2}}ds. \tag{16}$$

Integrating (16) from z to ∞, we find the following:

$$\varsigma''(z) \leq -(1-\tilde{y}_0)^{r_2/r_1}\varsigma^{r_2/r_1}(z)\int_z^\infty \left(\frac{1}{m(u)}\int_u^\infty \tilde{\omega}(s)\frac{\alpha^{r_2}(s)}{s^{r_2}}ds\right)^{1/r_1}du, \tag{17}$$

From the definition of $A(z)$, we see that $A(z) > 0$ for $z \geq z_1$, and using (13) and (17), we find the following:

$$\begin{aligned}A'(z) &= \frac{h'(z)}{h(z)}A(z) + h(z)\frac{\varsigma''(z)}{\varsigma(z)} - h(z)\left(\frac{\varsigma'(z)}{\varsigma(z)}\right)^2 \\ &\leq \frac{h'(z)}{h(z)}A(z) - \frac{1}{h(z)}A^2(z) \\ &\quad -(1-\tilde{y}_0)^{r_2/r_1}h(z)\varsigma^{r_2/r_1-1}(z)\int_z^\infty \left(\frac{1}{m(u)}\int_u^\infty \tilde{\omega}(s)\frac{\alpha^{r_2}(s)}{s^{r_2}}ds\right)^{1/r_1}du.\end{aligned}$$

Since $\varsigma'(z) > 0$, there exist $z_2 \geq z_1$ and $A_2 > 0$ such that the following holds:

$$\varsigma(z) > A_2.$$

Thus, we obtain the following:

$$A'(z) \leq -\Phi(z) + \frac{h'(z)}{h(z)}A(z) - \frac{1}{h(z)}A^2(z),$$

Thus, (7) holds. Proof of the theorem is completed. □

Definition 1. *Let*

$$D = \{(z,s) \in \mathbb{R}^2 : z \geq s \geq z_0\} \text{ and } D_0 = \{(z,s) \in \mathbb{R}^2 : z > s \geq z_0\}.$$

The function $G_i \in C(D, \mathbb{R})$ *fulfills the following conditions:*

(i) $G_i(z,s) = 0$ *for* $z \geq z_0$, $G_i(z,s) > 0$, $(z,s) \in D_0$;

(ii) The functions $h, v \in C^1([z_0, \infty), (0, \infty))$ and $g_i \in C(D_0, \mathbb{R})$ such that

$$\frac{\partial}{\partial s} G_1(z,s) + \frac{\beta'(s)}{\beta(s)} G_1(z,s) = g_1(z,s) G_1^{r_1/(r_1+1)}(z,s) \tag{18}$$

and

$$\frac{\partial}{\partial s} G_2(z,s) + \frac{h'(s)}{h(s)} G_2(z,s) = g_2(z,s) \sqrt{G_2(z,s)}. \tag{19}$$

Now, we present some Philos-type oscillation criteria for (1).

Theorem 1. Let (24) hold. If $\beta, h \in C^1([z_0, \infty), \mathbb{R})$ such that

$$\limsup_{z \to \infty} \frac{1}{G(z,z_1)} \int_{z_1}^{z} G(z,s) E_1(s) - \frac{g_1^{r_1+1}(z,s) G_1^{r_1}(z,s)}{(r_1+1)^{r_1+1}} \frac{2^{r_1} m(s) \beta(s)}{(\mu_1 s^2)^{r_1}} ds = \infty \tag{20}$$

for all $\mu_2 \in (0,1)$, and

$$\limsup_{z \to \infty} \frac{1}{G_2(z,z_1)} \int_{z_1}^{z} \left(G_2(z,s) \Phi(s) - \frac{h(s) g_2^2(z,s)}{4} \right) ds = \infty, \tag{21}$$

then (1) is oscillatory.

Proof. Let δ be a non-oscillatory solution of (1), we see that $\delta > 0$. Assume that (N_1) holds. Multiplying (6) by $G(z,s)$ and integrating the resulting inequality from z_1 to z, we obtain the following:

$$\int_{z_1}^{z} G(z,s) E_1(s) ds \leq B(z_1) G(z,z_1) + \int_{z_1}^{z} \left(\frac{\partial}{\partial s} G(z,s) + \frac{\beta'(s)}{\beta(s)} G(z,s) \right) B(s) ds$$
$$- \int_{z_1}^{z} \Theta(s) G(z,s) B^{\frac{r_1+1}{r_1}}(s) ds.$$

From (18), we obtain the following:

$$\int_{z_1}^{z} G(z,s) E_1(s) ds \leq B(z_1) G(z,z_1) + \int_{z_1}^{z} g_1(z,s) G_1^{r_1/(r_1+1)}(z,s) B(s) ds$$
$$- \int_{z_1}^{z} \Theta(s) G(z,s) B^{\frac{r_1+1}{r_1}}(s) ds. \tag{22}$$

Using Lemma 3 with $V = \Theta(s) G(z,s)$, $U = g_1(z,s) G_1^{r_1/(r_1+1)}(z,s)$ and $\delta = B(s)$, we obtain the following:

$$g_1(z,s) G_1^{r_1/(r_1+1)}(z,s) B(s) - \Theta(s) G(z,s) B^{\frac{r_1+1}{r_1}}(s)$$
$$\leq \frac{g_1^{r_1+1}(z,s) G_1^{r_1}(z,s)}{(r_1+1)^{r_1+1}} \frac{2^{r_1} m(z) \beta(z)}{(\mu_1 z^2)^{r_1}},$$

which, with (22) gives the following:

$$\frac{1}{G(z,z_1)} \int_{z_1}^{z} \left(G(z,s) E_1(s) - \frac{g_1^{r_1+1}(z,s) G_1^{r_1}(z,s)}{(r_1+1)^{r_1+1}} \frac{2^{r_1} m(s) \beta(s)}{(\mu_1 s^2)^{r_1}} \right) ds \leq B(z_1),$$

which contradicts (20).

Assume that (N_2) holds. Multiplying (7) by $G_2(z,s)$ and integrating the resulting inequality from z_1 to z, we find the following:

$$\int_{z_1}^{z} G_2(z,s)\Phi(s)ds \leq A(z_1)G_2(z,z_1)$$
$$+ \int_{z_1}^{z} \left(\frac{\partial}{\partial s}G_2(z,s) + \frac{h'(s)}{h(s)}G_2(z,s)\right)A(s)ds$$
$$- \int_{z_1}^{z} \frac{1}{h(s)}G_2(z,s)A^2(s)ds.$$

Thus,

$$\int_{z_1}^{z} G_2(z,s)\Phi(s)ds \leq A(z_1)G_2(z,z_1) + \int_{z_1}^{z} g_2(z,s)\sqrt{G_2(z,s)}A(s)ds$$
$$- \int_{z_1}^{z} \frac{1}{h(s)}G_2(z,s)A^2(s)ds$$
$$\leq A(z_1)G_2(z,z_1) + \int_{z_1}^{z} \frac{h(s)g_2^2(z,s)}{4}ds$$

and so

$$\frac{1}{G_2(z,z_1)}\int_{z_1}^{z}\left(G_2(z,s)\Phi(s) - \frac{h(s)g_2^2(z,s)}{4}\right)ds \leq A(z_1),$$

which contradicts (21). Proof of the theorem is completed. □

Corollary 1. *Let* (24) *hold. If* $\beta, h \in C^1([z_0, \infty), \mathbb{R})$ *such that*

$$\int_{z_0}^{\infty}\left(E_1(s) - \frac{2^{r_1}}{(r_1+1)^{r_1+1}}\frac{m(s)(\beta'(s))^{r_1+1}}{\mu_1^{r_1}s^{2r_1}\beta^{r_1}(s)}\right)ds = \infty \qquad (23)$$

and

$$\int_{z_0}^{\infty}\left(\Phi(s) - \frac{(h'(s))^2}{4h(s)}\right)ds = \infty, \qquad (24)$$

for some $\mu_1 \in (0,1)$ *and every* $A_1, A_2 > 0$, *then* (1) *is oscillatory.*

3. Example

This section presents some interesting examples to examine the applicability of theoretical outcomes.

Example 1. *Consider the following equation:*

$$\left(\delta + \frac{1}{2}\delta\left(\frac{1}{3}z\right)\right)^{(4)} + \frac{\tilde{\omega}_0}{z^4}\delta\left(\frac{1}{2}z\right) = 0, \ z \geq 1, \tilde{\omega}_0 > 0. \qquad (25)$$

Let $r_1 = r_2 = 1$, $m(z) = 1$, $\tilde{y}(z) = 1/2$, $\tilde{\alpha}(z) = z/3$, $\alpha(z) = z/2$ *and* $\tilde{\omega}(z) = \tilde{\omega}_0/z^4$. *Hence, it is easy to see that*

$$\int_{z_0}^{\infty} \frac{1}{m^{1/r_1}(s)}ds = \infty, \ E_1(z) = \frac{\tilde{\omega}_0}{16s}$$

and

$$\Phi(z) := \frac{\tilde{\omega}_0}{24}.$$

If we put $\beta(s) := z^3$ and $h(z) := z^2$, then we find the following:

$$\int_{z_0}^{\infty} \left(E_1(s) - \frac{2^{r_1}}{(r_1+1)^{r_1+1}} \frac{m(s)(\beta'(s))^{r_1+1}}{\mu_1^{r_1} s^{2r_1} \beta^{r_1}(s)} \right) ds$$
$$= \int_{z_0}^{\infty} \left(\frac{\tilde{\omega}_0}{16s} - \frac{9}{2\mu_1 s} \right) ds$$

and

$$\int_{z_0}^{\infty} \left(\Phi(s) - \frac{(h'(s))^2}{4h(s)} \right) ds$$
$$= \int_{z_0}^{\infty} \left(\frac{\tilde{\omega}_0}{24} - 1 \right) ds.$$

Thus,

$$\tilde{\omega}_0 > 72 \tag{26}$$

and

$$\tilde{\omega}_0 > 24. \tag{27}$$

From Corollary 1, Equation (25) is oscillatory if $\tilde{\omega}_0 > 72$.

Example 2. Consider the following equation:

$$\left(z(\delta + \tilde{y}_0 \delta(\gamma z))''' \right)' + \frac{\tilde{\omega}_0}{z^3} \delta(\eta z) = 0, \ z \geq 1, \tag{28}$$

where $\tilde{y}_0 \in [0,1)$, $\gamma, \eta \in (0,1)$ and $\tilde{\omega}_0 > 0$. Let $r_1 = r_2 = 1$, $m(z) = z$, $\tilde{y}(z) = \tilde{y}_0$, $\tilde{\alpha}(z) = \gamma z$, $\alpha(z) = \eta z$ and $\tilde{\omega}(z) = \tilde{\omega}_0/z^3$. Hence, if we set $\beta(s) := z^2$ and $h(z) := z$, then we get

$$E_1(z) = \frac{\tilde{\omega}_0(1-\tilde{y}_0)\eta^3}{z}, \ \Phi(z) = \frac{\tilde{\omega}_0(1-\tilde{y}_0)\eta}{4z}.$$

Thus, (23) and (24) become the following:

$$\int_{z_0}^{\infty} \left(E_1(s) - \frac{2^{r_1}}{(r_1+1)^{r_1+1}} \frac{m(s)(\beta'(s))^{r_1+1}}{\mu_1^{r_1} s^{2r_1} \beta^{r_1}(s)} \right) ds$$
$$= \int_{z_0}^{\infty} \left(\frac{\tilde{\omega}_0(1-\tilde{y}_0)\eta^3}{s} - \frac{2}{\mu_1 s} \right) ds$$

and

$$\int_{z_0}^{\infty} \left(\Phi(s) - \frac{(h'(s))^2}{4h(s)} \right) ds$$
$$= \int_{z_0}^{\infty} \left(\frac{\tilde{\omega}_0(1-\tilde{y}_0)\eta}{4s} - \frac{1}{4s} \right) ds.$$

So,

$$\tilde{\omega}_0 > \frac{2}{(1-\tilde{y}_0)\eta^3} \tag{29}$$

and

$$\tilde{\omega}_0 > \frac{1}{(1-\tilde{y}_0)\eta}.$$

From Corollary 1, Equation (28) is oscillatory if (29) holds.

4. Conclusions

In this work, we prove some new oscillation theorems for (1). New oscillation results are established that complement related contributions to the subject. We used the Riccati technique and integral averages technique to obtain some new results to the oscillation of Equation (1) under the condition $\int_{z_0}^{\infty} \frac{1}{m^{1/r_1}(s)} ds = \infty$. In future work, we will study this type of equation under the following condition:

$$\int_{z_0}^{\infty} \frac{1}{m^{1/r_1}(s)} ds < \infty,$$

We also introduce some important oscillation criteria of differential equations of the fourth-order and under the following:

$$\varsigma(z) = \delta(z) + \tilde{y}(z) \sum_{i=1}^{j} \delta_i(\tilde{\alpha}(z)).$$

Author Contributions: Conceptualization, A.A., O.B. and R.A.E.-N.; methodology, A.A., O.B. and R.A.E.-N.; investigation, A.A., O.B. and R.A.E.-N.; resources, A.A., O.B. and R.A.E.-N.; data curation, A.A., O.B. and R.A.E.-N.; writing—original draft preparation, A.A., O.B. and R.A.E.-N.; writing—review and editing, A.A., O.B. and R.A.E.-N.; supervision, A.A., O.B. and R.A.E.-N.; project administration, A.A., O.B. and R.A.E.-N.; funding acquisition, A.A., O.B. and R.A.E.-N. All authors have read and agreed to the published version of the manuscript.

Funding: This research received no external funding.

Institutional Review Board Statement: Not applicable.

Informed Consent Statement: Not applicable.

Data Availability Statement: Not applicable.

Conflicts of Interest: The authors declare no conflict of interest.

References

1. Althobati, S.; Bazighifan, O.; Yavuz, M. Some Important Criteria for Oscillation of Non-Linear Differential Equations with Middle Term. *Mathematics* **2021**, *9*, 346. [CrossRef]
2. Hale, J.K. *Theory of Functional Differential Equations*; Springer: New York, NY, USA, 1977.
3. Agarwal, R.P.; Bohner, M.; Li, T.; Zhang, C. A new approach in the study of oscillatory behavior of even-order neutral delay diferential equations. *Appl. Math. Comput.* **2013**, *225*, 787–794.
4. Baculikova, B.; Dzurina, J.; Li, T. Oscillation results for even-order quasi linear neutral functional differential equations. *Electron. J. Differ. Equ.* **2011**, *2011*, 1–9.
5. Agarwal, R.P.; Bazighifan, O.; Ragusa, M.A. Nonlinear Neutral Delay Differential Equations of Fourth-Order: Oscillation of Solutions. *Entropy* **2021**, *23*, 129. [CrossRef] [PubMed]
6. Xing, G.; Li, T.; Zhang, C. Oscillation of higher-order quasi linear neutral differential equations. *Adv. Differ. Equ.* **2011**, *2011*, 1–10. [CrossRef]
7. Bazighifan, O.; Alotaibi, H.; Mousa, A.A.A. Neutral Delay Differential Equations: Oscillation Conditions for the Solutions. *Symmetry* **2021**, *13*, 101. [CrossRef]
8. Zhang, C.; Li, T.; Saker, S. Oscillation of fourth-order delay differential equations. *J. Math. Sci.* **2014**, *201*, 296–308. [CrossRef]
9. Chatzarakis, G.E.; Elabbasy, E.M.; Bazighifan, O. An oscillation criterion in 4th-order neutral differential equations with a continuously distributed delay. *Adv. Differ. Equ.* **2019**, *336*, 1–9.
10. Kiguradze, I.T.; Chanturiya, T.A. *Asymptotic Properties of Solutions of Nonautonomous Ordinary Differential Equations*; Kluwer Academic Publishers: Dordrecht, The Netherlands, 1993.
11. Moaaz, O.; Awrejcewicz, J.; Bazighifan, O. A New Approach in the Study of Oscillation Criteria of Even-Order Neutral Differential Equations. *Mathematics* **2020**, *12*, 197. [CrossRef]
12. Elabbasy, E.M.; Cesarano, C.; Bazighifan, O.; Moaaz, O. Asymptotic and oscillatory behavior of solutions of a class of higher order differential equation. *Symmetry* **2019**, *11*, 1434. [CrossRef]
13. Philos, C. A new criterion for the oscillatory and asymptotic behavior of delay differential equations. *Bull. Acad. Pol. Sci. Sér. Sci. Math.* **1981**, *39*, 61–64.

Article

BVPs Codes for Solving Optimal Control Problems

Francesca Mazzia [1,*,†] **and Giuseppina Settanni** [2,†]

1. Dipartimento di Informatica, Università degli Studi di Bari Aldo Moro, 70125 Bari, Italy
2. Dyrecta Laboratory, Instituto di Ricerca, Via Vescovo Simplicio 45, 70014 Conversano, Italy; giuseppina.settanni@dyrecta.com
* Correspondence: francesca.mazzia@uniba.it
† These authors contributed equally to this work.

Abstract: Optimal control problems arise in many applications and need suitable numerical methods to obtain a solution. The indirect methods are an interesting class of methods based on the Pontryagin's minimum principle that generates Hamiltonian Boundary Value Problems (BVPs). In this paper, we review some general-purpose codes for the solution of BVPs and we show their efficiency in solving some challenging optimal control problems.

Keywords: optimal control; indirect methods; boundary value problems

Citation: Mazzia, F.; Settanni, G. BVPs Codes for Solving Optimal Control Problems. *Mathematics* **2021**, *9*, 2618. https://doi.org/10.3390/math9202618

Academic Editors: Fasma Diele and Janusz Brzdek

Received: 30 June 2021
Accepted: 12 October 2021
Published: 17 October 2021

Publisher's Note: MDPI stays neutral with regard to jurisdictional claims in published maps and institutional affiliations.

Copyright: © 2021 by the authors. Licensee MDPI, Basel, Switzerland. This article is an open access article distributed under the terms and conditions of the Creative Commons Attribution (CC BY) license (https://creativecommons.org/licenses/by/4.0/).

1. Introduction

Many optimal control problems arise from an interest in observing the dynamic behavior of a state variable described by a dynamic equation, namely by a differential equation, in several areas of applications such as biology, chemistry, economy, physics, and engineering. For example, we can consider the development of a specific species of animals in an ecological preserve, the dynamical behavior of a chemical process, the evolution of the selling trend of a company, or the simulation of high-performance racing vehicles. The dynamical behavior of this kind of problems is influenced by the choice of control variables, as it might be incorporating the presence of predators in the ecological preserve, moreover, both state and control variables must fulfil constraints, and minimize or maximize an objective function.

Numerical methods solving optimal control problems were considered starting from the 1950s, when Bellman introduced the dynamic programming [1], that requires solving a partial differential equation, called the Hamiltonian-Jacobi-Bellman equation. Through time the numerical approaches can be mainly divided into two classes: direct methods and indirect methods [2,3]. Perhaps the first class of direct methods is the most widely applied, it transforms the problem into a nonlinear optimization problem or nonlinear programming problem, essentially this class is focused on the use of optimization techniques. The second class of the indirect methods transforms the original optimal control problem into a two-point boundary value problem, highlighting particular attention to numerical methods solving differential equation systems. The last strategy is often considered disadvantageous for figuring out challenging optimal control problem.

As against this last opinion, this work aims to review many of the available general-purpose codes solving boundary value problems, able to figure out optimal control problems arising from adopting an indirect approach. The review is also devoted to some numerical strategies that are useful and sometime necessary to numerically solve the problem, such as continuation techniques associated with suitable penalty functions.

The most used solver for indirect methods has been the shooting method, based on guessing the value of the unknown boundary condition at one end of the interval, so that an initial value problem is solved to obtain the solution at the other end of the interval that is already known. Although the shooting method is simple to apply, it is not particularly advantageous to use when the boundary value problem is ill conditioned or stiff, and purposely when the optimal control problem is hypersensitive [4]. To overcome this matter the multiple shooting method is considered, specifically the time interval is partitioned in more subintervals and the shooting method is applied over each of these intervals. Another class of methods widely used, since it is the most robust and fast converging, is the class of collocation methods, where piecewise polynomials are used to parametrize the state and control variables. Finally, the solution is computed solving a nonlinear system by means of root finding techniques.

In the literature there exist different general-purpose open-source codes solving boundary value problems that are highly suitable in solving stiff and singular perturbations problems. Many of them have been implemented in Fortran, which has been for many years the preferred language for scientific computing. Some effort has been however accomplished to make them also available in problem-solving environments such as Matlab or R. The first bvp codes, as colsys/colnew [5,6], twpbvp [7], twpbvpl [8], acdc and colmod [9], coldae [10], mirkdc [11] and BVP_M-2 [12,13] have been written in Fortran/Fortran90. A collection of the last releases of many of the cited Fortran codes, together with the driver that allows a common input definition, and a list of numerical examples arising in several applications are available in the web site Test set for BVP solvers [14,15].

The Matlab environment allows the use of two functions, named bvp4c [16] and bvp5c [17], for solving BVPs. Other interesting codes that are usable in Matlab are bvptwp [18], TOM [19], HOFiD_bvp [20] and bvpSuite2.0 [21], based on the code sbvp [22] for the solution of singular problems. The code bvpSuite2.0 could be used also for singular BVPs and differential algebraic problems of index 1. For the R community is instead available the package called bvpSolve that allows the running in R of many of the available Fortran codes [4,23]. In Python the package scipy.integrate includes the function solve_bvp [24], a routine based on BVP_M-2 and similar to the bvp4c Matlab code. All of them solve two-point boundary value problems, this means that applied to a second order boundary value problems, they transform the original problem into a system of first-order differential equations with boundary conditions, except for the collocation codes colsys, colnew, colmod, coldae, bvpSuite, and the high-order finite difference code HOFiD_bvp, since each of them can be applied directly to higher order problems.

Our aim is to apply some of the cited codes for figuring out boundary value problems coming up using indirect methods to optimal control problems. Meanwhile, we will highlight some matters that can arise in handling bvp solvers, such as the choice of an initial mesh, or the use of a continuation technique for nonlinear problems. To this aim we show by some test problems how the proper use of these techniques and a good choice of the input parameters can allow us to obtain a solution in more efficient way than we could achieve using default parameters. Since the aim of this paper is not to make a comparison between the selected codes, we do not show the execution time, but we point out how the choice of a code depends on the problem.

We use as platform to run the experiment the Matlab environment and we consider the codes available in the Matlab distribution, bvptwp and TOM. We do not present the results for the collocation code bvpSuite2.0 because it does not give in output the same information of the other codes and it does not allow using a numerical Jacobian. For R-users all the examples could be solved using all the codes available in the bvpSolve package. Since bvpSolve run the Fortran codes by means of an interface, the results are the same obtained by the original Fortran codes.

The paper shows a list of a few interesting problems, for other applications of the same codes, here considered, to more involved optimal control problems, we refer the reader to [25–28]. Moreover, we highlight that it is not our aim to compare direct and indirect methods, but only to show the efficiency of indirect methods that often are not taken into consideration because users do not know the potentiality of general-purpose codes for BVPs.

The paper is organized as follows: in Section 2 we briefly introduce the indirect methods; in Section 3 we review codes for solving boundary value problems (BVPs) that are illustrated and classified through different programming environments, in particular Fortran codes are allocated in Section 3.1, Matlab codes in Section 3.2 and R codes in Section 3.3. Finally, in Sections 4–8 interesting optimal control problems are solved using indirect methods and the BVPs related. In Section 9 we give some conclusions, highlighting the potentiality of the BVP codes considered.

2. Optimal Control Problems: Indirect Methods

Given a non-empty compact time interval $[t_0, t_f] \subset \mathcal{R}$, with $t_0 < t_f$, an optimal control problem is defined as

$$\begin{aligned} \text{minimize} \quad & \varphi(t_f, \mathbf{x}_f) + \int_{t_0}^{t_f} L(t, \mathbf{x}, \mathbf{u}) \, dt, \\ & \mathbf{x}' = f(t, \mathbf{x}, \mathbf{u}), \\ & \mathbf{b}(\mathbf{x}(t_0), \mathbf{x}(t_f)) = \mathbf{0}, \\ & \mathbf{u} \in \mathcal{U}, \end{aligned} \quad (1)$$

where φ and L are sufficiently smooth functions involved in the minimization of the objective function, $\mathbf{x}(t) \in \mathbb{R}^n$ is the state variable of the dynamical system, $\mathbf{u}(t) \in \mathcal{U} \subset \mathbb{R}^m$ is the control variable and $\mathcal{U}$ the set of admissible controls, f is a regular function and $\mathbf{b}(\mathbf{x}(t_0), \mathbf{x}(t_f)) = \mathbf{0}$ are the general boundary conditions. Furthermore, Problem (1) might be subject to a path constraint that can be expressed by a mixed control-state constraint $\mathbf{c}(t, \mathbf{x}, \mathbf{u}) \leq \mathbf{0}$ or a pure state constraint $s(t, \mathbf{x}) \leq 0$.

There exist two main approaches solving optimal control problems (1), direct methods and indirect methods [2,29]. Direct methods suitably discretize an infinite-dimensional optimal control problem, giving back a finite-dimensional optimization problem that can be solved using appropriate nonlinear programming methods, such as sequential quadratic programming. This approach results robust and efficient if applied to several problems, besides not requiring a strong knowledge in optimal control theory, it becomes highly advantageous to use.

On the other hand, indirect methods are instead related to the Pontryagin's minimum principle [29], a necessary condition for optimality that transforms the original Problem (1) into a two-point boundary value problem for state and adjoint Lagrange multiplier functions, defined as

$$\begin{aligned} & \mathbf{x}' = f(t, \mathbf{x}, \mathbf{u}), \\ & \boldsymbol{\lambda}' = -H_{\mathbf{x}}(t, \mathbf{x}, \mathbf{u}, \boldsymbol{\lambda}), \\ & \mathbf{b}(\mathbf{x}(t_0), \mathbf{x}(t_f)) = \mathbf{0}, \\ & \mathbf{b}_{\mathbf{x}(t_0)}(\mathbf{x}(t_0), \mathbf{x}(t_f))\omega = \boldsymbol{\lambda}(t_0), \\ & \mathbf{b}_{\mathbf{x}(t_f)}(\mathbf{x}(t_0), \mathbf{x}(t_f))\omega = -\varphi_{\mathbf{x}}(t_f, \mathbf{x}_f) - \boldsymbol{\lambda}(t_f), \end{aligned} \quad (2)$$

where $H(t, \mathbf{x}, \mathbf{u}, \boldsymbol{\lambda}) = L(t, \mathbf{x}, \mathbf{u}) + \boldsymbol{\lambda} \cdot f(t, \mathbf{x}, \mathbf{u})$ is the Hamiltonian function and the optimal control $\mathbf{u}^*(t)$ is obtained by a local optimization of the Hamiltonian, namely $\mathbf{u}^*(t) = \arg\min_{\mathbf{u} \in \mathcal{U}} H(t, \mathbf{x}, \mathbf{u}, \boldsymbol{\lambda})$. Pro this approach there is the possibility to compute an accurate numerical solution; however, against we find some drawbacks, such as the necessity to have a good initial guess for the solution of the generated nonlinear boundary value problem. Now, to overcome this matter we focus on the application of some well-known two-point boundary value codes that are considered extremely efficient and robust to solve the BVP (2).

3. Codes for BVPs

Boundary value problems arise in many fields of application, so in the last 40 years a great effort has been done to develop efficient methods solving this kind of problems.

Among them many are methods applied to two-point boundary value problems, i.e., to systems of first-order ordinary differential equations with boundary conditions, others can be applied directly to second or high-order boundary value problems without any transformation of the original problem. Moreover, these codes are available in different programming environment, so in the following we will give information about their characteristics.

3.1. Fortran Codes

The code colsys was written by U. Ascher, R. Matteij and R. Russell [5] and it is based on method of spline collocation at Gaussian points and solves mixed-order systems of multipoint BVPs, high-order equations, problems with non-separated boundary conditions and problems with singularity. The code computes the solution on a sequence of meshes that are refined using the equidistribution of error to satisfy the required input tolerance. The error estimate is obtained roughly at each step halving the mesh. The components of the collocation solution are expressed by B-spline basis, which are evaluated by the de Boor's algorithms. Indeed, the damped Newton's method of quasilinearization is used for solving the nonlinear problems.

The code colnew [6,30] is the descendant of colsys and, contrary to this last, it uses a Runge–Kutta monomial representation for the piecewise polynomial solution, instead of B-spline basis. This change returns a code faster than the native version colsys.

The codes twpbvp, twpbvpl and acdc were written by J. R. Cash and his collaborators. The code twpbvp [7], differently from colsys, uses mono-implicit Runge–Kutta formulae and a deferred correction method for solving two-point boundary value problems. The mono-implicit Runge–Kutta formulae are implemented applying the deferred correction procedure, which allows discovery of the solution of a high-order method using only low order schemes. The code guarantees to construct a mesh refinement that is very suitable for singular perturbation problems.

The code twpbvpl, differently from twpbvp, is based on three Lobatto Runge–Kutta formulae of order 4, 6, 8, which are implemented using a suitable deferred correction scheme, solved with a damped Newton iteration scheme. The code is devoted in solving efficiently nonlinear stiff two-point boundary value problems.

The code acdc [9] has been developed from twpbvpl including an automatic continuation strategy, implemented to suitably solve linear and nonlinear singular perturbation problems characterized from a small parameter ϵ. The parameter ϵ often brings about stiffness in the problem, so that for a nonlinear problem a good initial solution is required to reach the convergence of the Newton method. The continuation strategy arises to overcome these matters, specifically it consists of selecting an initial perturbation parameter ϵ_0, chosen to compute a solution of a problem not particularly stiff, usually for $\epsilon_0 \approx 1$, and satisfying a certain exit tolerance tol. The idea is to obtain an initial rough profile of the solution of the problem for a desired perturbation parameter ϵ. Then, chosen an integer N_ϵ the interval $[\epsilon_0, \epsilon]$ is discretized in N_ϵ subintervals, so that

$$\epsilon_0 > \epsilon_1 > \cdots > \cdots > \epsilon_{N_\epsilon - 1} > \epsilon_{N_\epsilon}.$$

Now, $N_\epsilon + 1$ boundary value problems satisfying an exit tolerance tol are iteratively computed, so that the solution of the problem obtained at iteration $i = 0, \ldots, N_\epsilon - 1$, for ϵ_i on a mesh π_i, is the initial solution of the next problem with perturbation parameter ϵ_{i+1}. A crucial point of this strategy is the selection of the initial parameter ϵ_0 and the value of discretizazion N_ϵ, both depend on the problem. In codes such as acdc ϵ_0 is set equal to 0.5 by default; however the suggestion is to consider ϵ_0 as a value not extremely small allowing the obtaining of an accurate solution of the problem for that value of perturbation; The code acdc chooses the sequences of parameters and the total number of continuation steps automatically. It is however possible to implement a continuation strategy for the other codes, in this case for N_ϵ it would be convenient to start with a small integer and then double or increment it, if the procedure does not converge.

The code `colmod` [9] is a modified version of the code `colsys` using the same continuation strategy adopted in acdc.

The codes `twpbvpc`, `twpbvplc` and `acdcc` [14] are the modified version of the codes `twpbvp`, `twpbvpl` and `acdc` that implement a mesh selection strategy based on the estimation of the local error and of two conditioning parameters [31]. This hybrid mesh strategy has first been used in the Matlab code TOM, described in the next section.

The code `mirkdc` written by W. Enright and P. Muir [11] uses MIRK method and controls the defect, also BVP_M-2 written by J.J. Boisvert, P. Muir and R. Spiteri [12] is based on MIRK methods, but this last controls both the defect and/or the global error, giving, moreover, information about the conditioning constant.

Detailed information about all the numerical schemes and techniques related to the Fortran codes in this subsection can be found in [32] where a review of global methods for solving BVPs is presented.

3.2. Matlab Codes

The BVP codes available officially in the Matlab environment are `bvp4c` [33] and `bvp5c` [34]. The code `bvp4c` [16] is based on a collocation method with a C^1 piecewise cubic polynomial, or equivalently on an implicit Runge–Kutta formula with a continuous extension, namely the collocation method is equivalent to a three-stages Lobatto IIIa implicit Runge–Kutta formula. This code implements a method of order four and solves a large class of BVP, such as equations with non-separated boundary conditions, singular problems, Sturm–Liouville problems. An advantage of this code is being able to compute numerical partial derivatives and use a vectorized finite difference Jacobian. Differently from the other codes the error estimation and the mesh selection are based on the residual estimation. We recall that if $S(x)$ approximates the solution $y(x)$, then the residual control in the differential equation $y'(x) = f(x, y(x))$ is given by $r(x) = |S'(x) - f(x, S(x))|$.

The code `bvp5c` is based on the four-stages Lobatto IIIa formula, giving a method of order five. Contrarily to `bvp4c`, `bvp5c` controls the residual and the true approximate error. It is clear that if the BVP is well-conditioned a small residual implies a small true error, but this is not satisfied if the BVP is ill-conditioned, hence the strategy to control the residual and the true error is more efficient than the one applied in `bvp4c`.

The next two codes TOM and HOFiD_bvp belong to the class of Boundary Value Methods [35], especially suitable for solving BVPs.

The code TOM [19], based on the TOP Order Methods and the BS method of order four, six, eight and ten distinguishes for the use of conditioning in the mesh selection strategy. In [36] the authors analyzed how the conditioning and the stiffness of a problem depend on the estimation of the following conditioning parameters:

- κ conditioning constant with respect to all type of perturbation, computed using the maximum norm;
- κ_1 conditioning constant with respect to a perturbation of the boundary conditions, computed using the maximum norm;
- κ_2 conditioning constant with respect to a perturbation of the differential problem, computed using the maximum norm;
- γ_1 conditioning constant with respect to a perturbation of the boundary conditions, computed using the one norm;
- σ the stiffness ratio.

Specifically, the problem is: well-conditioned if κ, κ_1, γ_1 and σ are of moderate size; stiff if $\sigma \gg 1$; ill-conditioned if $\kappa \gg 1$ and $\gamma \gg 1$; ill posed if $\kappa_2 > \kappa_1$. A complete description of the parameters and the algorithms used to compute their approximation is presented in [37]. The hybrid mesh selection algorithm controls the approximation of conditioning parameters and chooses the mesh points to have an estimation of those discrete quantities close to the continuous ones. Meanwhile, the code controls that the error of the solution computed is less than a prescribed tolerance. The error approximation is computed using a deferred correction technique with a higher order method, moreover a quasi-linearization technique is implemented to solve nonlinear problem. The release of May 2021, which has been used for the numerical tests in this paper, has the possibility to choose two different mesh selections, one suitable for regular problem and the other one for stiff or singular perturbation problems.

The code HOFiD_bvp [20] is based on high-order finite difference schemes (HOFiD) of order four, six, eight and ten, and an upwind method. Each derivative in the high-order boundary value problem is approximated directly by these schemes, hence it is not required any transformation of the problem in a system of first-order differential equations. The error estimation is computed applying the deferred correction technique to two consecutive order methods. The mesh selection is based on the error equidistribution. For nonlinear problems, the code uses a continuation strategy, as explained previously, and also combines an order variation strategy, this means that a solution of the problem obtained with a lower order and tolerance can be considered to be initial solution to run the code with higher order and tolerance. The strategy adopted returns a code suitable to solve high-order boundary value problems that can be singularly perturbed, singular, with discontinuous terms and multipoint. Other versions of the code solve singular second order initial value problems [38], Sturm–Liouville problems [39] and multi-parameters spectral problems [40].

An interesting code for solving high-order BVPs is the bvpSuite2.0 package, based on collocation methods. The collocation points could be chosen by the users among Gauss, Lobatto, uniform or user defined points. The code solves implicit BVPs, eigenvalue problems, differential algebraic problems of index 1 and it is particularly suited for singular problems. BvpSuite2.0 [21] is the evolution of two previous versions of the code with improved usability. The mesh selection strategy used is described in [41].

Finally, we consider the Matlab code bvptwp [18] based on an efficient translation of the Fortran codes twpbvp, twpbvpl and acdc in the Matlab environment, which are named twpbvp_m, twpbvp_l, acdc. Moreover, the Matlab package also contains the translation of the Fortran version of the same codes that use a hybrid mesh selection based on conditioning, similar to the one used in the code TOM, called twpbvpc_m, twpbvpc_l, acdcc. The code bvptwp is available on the calgo website and on the web-page called Test Set for BVP Solvers [15]. The version used in this paper is the release of May 2021.

3.3. R Codes

In recent years, the use of the open-source software R is upward among the problem-solving environments (PSEs) , and although it is mainly used as a software for statistics and visualization, several powerful methods solving differential equations have been developed. In this regard we highlight the package bvpSolve [23], which, using an interface, implements all the Fortran codes introduced in Section 3.1.

3.4. Experiments

Since our aim is to show the suitability and the efficiency of the BVP solvers in computing the solution of the Hamiltonian boundary value problems deriving from the application of the indirect method to optimal control problems, in the following sections we carry out some interesting numerical tests. We run experiments using the Matlab codes bvp4c, bvp5c, and bvptwp. For the last solver we consider all the codes available, i.e., twpbvp_m, twpbvp_l, twpbvpc_m, twpbvpc_l, acdc, acdcc. We also add the results obtained with the new release of the code TOM (May 2021). This code allows the choice of a boundary value method of specific order and a mesh variation strategy. For all the examples we choose the BS method of order 4 and we denote by tom the code run using a mesh variation for regular problems and by tomc the one implementing a mesh variation suited for stiff problems. For R-users all the examples could be solved applying all the codes included in the bvpSolve package. Since bvpSolve runs the Fortran codes by an interface, the obtained results are similar to those computed by means of the original Fortran codes. We also observe that some of the codes considered here for the numerical tests are also present in the R package bvpSolve rel. 1.4.2. The R version of these codes on the same examples show comparable results.

In our tests we use an initial mesh with 16 equidistant points and an initial solution with zero elements, except in some examples where specified. Moreover, the maximal mesh allowed has been set to 10^4 and the function evaluations have been vectorized. In the tables we report the number of points in the final mesh fM (in reading this value we recall that the code TOM does not use any auxiliary steps but all the others codes needs also several intermediate steps depending on the order of the methods used), the total number of vectorized function evaluation NVF and the mixed relative error on some significant components of the solution defined for a generic component x by the following formula

$$\max_i \frac{|x_i - x(t_i)|}{(1 + |x(t_i)|)}$$

where x_i is the numerical approximation of $x(t_i)$. If the exact solution of the test problem is not available, the error is computed by running the code twpbvpc_l using a doubled mesh and a halved input tolerance. For all the codes we give in input equal absolute and relative tolerances. If the codes twpbvp_m/twpbvpc_m, twpbvp_l/twpbvpc_l, acdc/acdcc give the same results we report only one result in the tables. If a code cannot solve the problem, we put * in the tables.

4. Hypersensitive Optimal Control Problems

The first class of examples we consider is the class of hypersensitive optimal control problems. Problems in this class are stiff, and need a suitable mesh variation strategy when solved using both direct and indirect methods. Usually, they are considered extremely difficult to be solved by indirect methods, because the solution is sensitive to changes in the initial conditions. In [42] the authors describe a dichotomic basis method which is inspired to the computation of the solution of singular perturbation problems for stiff initial value problems. In the following examples we show that general-purpose finite differences codes can solve very efficiently this class of problems. The codes can be applied for the numerical solution of completely hypersensitive problems whose solution has fast rates in all directions and partially hypersensitive problems, with the fast rate in only one direction.

4.1. Nonlinear Mass Spring System with Quadratic Cost

As first example we consider a hypersensitive nonlinear mass spring system [43], where the mass position x is defined such that the spring is unstretched when $x = 0$. The spring force is $F_s(x) = -k_1 x - k_2 x^3$. The control is exerted on the mass by an external force denoted by $F(t)$, hence the control input is $u(t) = F(t)$. The equation of motion of the mass is $mx'' = F_s(x) + F(t)$. We assume that $k_1 = 1$, $k_2 = 1$ and $m = 1$.

The optimal control problem needs to determine the control u on the fixed time interval $[0, T]$ such that

$$\begin{cases} \min_{x,u} \frac{1}{2} \int_0^T (x^2 + v^2 + u^2) \, dt \\ x' = v \\ v' = -x - x^3 + u \\ x(0) = 1, \ v(0) = 0, \ x(T) = 0.75, \ v(T) = 0. \end{cases}$$

The associated Hamiltonian is

$$H(x, v, \lambda, \mu, u) = \frac{1}{2}(x^2 + v^2 + u^2) + \lambda v + \mu(-x - x^3 + u)$$

and the optimal control, obtained by computing $\frac{\partial H}{\partial u} = 0$, is given by $u^* = -\mu$. Therefore, applying the indirect method the optimal control problem is equivalent to solve the following BVP

$$x' = v$$
$$v' = -x - x^3 - \mu$$
$$\lambda' = -x + \mu(1 + 3x^2) \quad (3)$$
$$\mu' = -v - \lambda$$
$$x(0) = 1, \; v(0) = 0, \; x(T) = 0.75, \; v(T) = 0.$$

In Figure 1 we show the solution for $T = 20$ and $T = 40$. In Table 1 we present some results obtained increasing the value of T from 20 to $T = 2 \times 10^6$. First, we choose an initial mesh of 16 equidistant points and try to run all the codes, except the codes acdc and acdcc, since for this formulation of the problem there is not a parameter to be used for continuation. If on one hand, for $T = 20$ all the methods converge to the solution, and for $T = 2 \times 10^4$ only the codes bvp4c and bvp5 fail, on the other hand for $T = 2 \times 10^6$ no one goes to convergence except the codes tom and tomc (see Table 2). Essentially, there are some troubles with a singular Jacobian for bvp4c and bvp5c, or a drawback with the maximum number of mesh points allowed with the other codes. In the last case we could increase the maximum value of mesh points; however, we will try to differently overcome this matter and to debunk the idea that the indirect methods are not as competitive as direct ones.

Table 1. Nonlinear Mass spring: final mesh (fM), total number of vectorized function evaluation (NVF) and mixed errors for x, v, u. The solution is computed starting from an initial mesh with 16 equidistant points.

				$tol = 10^{-4}$						
			$T = 20$					$T = 2 \times 10^4$		
	fM	NVF	Error x	Error v	Error u	fM	NVF	Error x	Error v	Error u
bvp4c	71	35	6.0×10^{-6}	9.8×10^{-6}	1.7×10^{-5}	*	*	*	*	*
bvp5c	294	1001	9.0×10^{-9}	2.0×10^{-8}	4.7×10^{-8}	*	*	*	*	*
twpbvp_m	23	52	2.5×10^{-6}	4.4×10^{-6}	4.8×10^{-6}	158	263	3.8×10^{-6}	2.4×10^{-6}	3.0×10^{-7}
twpbvpc_m	38	52	2.4×10^{-6}	4.5×10^{-6}	5.1×10^{-6}	201	246	1.2×10^{-6}	1.7×10^{-6}	2.3×10^{-6}
twpbvp_l	27	54	1.7×10^{-6}	2.1×10^{-6}	4.4×10^{-6}	97	200	6.9×10^{-6}	1.1×10^{-5}	3.2×10^{-5}
twpbvpc_l	27	54	1.7×10^{-6}	2.1×10^{-6}	4.4×10^{-6}	104	246	5.7×10^{-6}	6.2×10^{-6}	2.8×10^{-5}
tom	116	14	2.8×10^{-6}	1.7×10^{-6}	3.1×10^{-6}	426	30	7.7×10^{-6}	4.5×10^{-6}	7.6×10^{-6}
tomc	136	16	1.2×10^{-6}	1.6×10^{-6}	2.4×10^{-6}	526	41	7.5×10^{-7}	1.1×10^{-6}	1.5×10^{-6}
				$tol = 10^{-6}$						
bvp4c	254	49	2.4×10^{-8}	6.6×10^{-8}	1.5×10^{-7}	*	*	*	*	*
bvp5c	392	1221	1.2×10^{-10}	1.2×10^{-10}	1.7×10^{-10}	*	*	*	*	*
twpbvp_m	42	50	1.1×10^{-8}	1.3×10^{-8}	2.3×10^{-8}	254	271	1.3×10^{-7}	1.0×10^{-7}	1.1×10^{-7}
twpbvpc_m	57	73	3.2×10^{-8}	4.4×10^{-8}	5.0×10^{-8}	306	306	8.1×10^{-7}	7.1×10^{-7}	6.3×10^{-7}
twpbvp_l	48	78	2.0×10^{-8}	1.9×10^{-8}	2.3×10^{-8}	152	207	1.7×10^{-8}	2.2×10^{-8}	2.5×10^{-8}
twpbvpc_l	58	78	2.0×10^{-8}	1.9×10^{-8}	2.3×10^{-8}	136	253	1.7×10^{-8}	2.2×10^{-8}	2.5×10^{-8}
tom	196	19	1.6×10^{-7}	2.2×10^{-7}	2.8×10^{-7}	481	33	4.5×10^{-7}	4.1×10^{-7}	5.1×10^{-7}
tomc	166	17	7.2×10^{-7}	6.8×10^{-7}	7.4×10^{-7}	511	44	2.1×10^{-7}	2.7×10^{-7}	3.1×10^{-7}

Table 2. Nonlinear Mass spring, $T = 2 \times 10^6$: final mesh (fM), total number of vectorized function evaluation (NVF) and mixed errors for x, v, u, initial mesh with 16 equidistant points.

	fM	NVF	Error x	Error v	Error u	fVM	NVF	Error x	Error v	Error u
tom	6266	256	6.4×10^{-7}	8.9×10^{-7}	1.1×10^{-6}	6266	256	6.4×10^{-7}	8.9×10^{-7}	1.1×10^{-6}
tomc	1291	177	6.6×10^{-7}	7.5×10^{-7}	1.4×10^{-6}	1236	180	1.7×10^{-7}	1.6×10^{-7}	1.8×10^{-7}

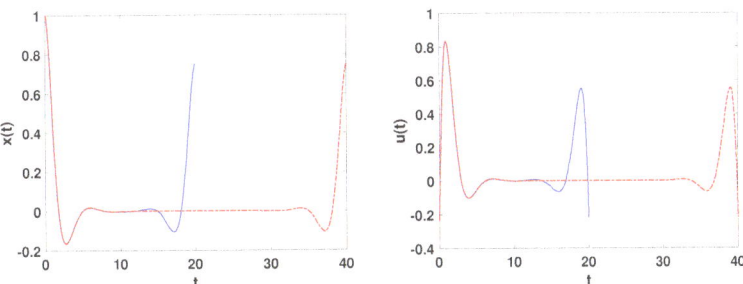

Figure 1. Mass spring: solution in time for the mass position x on the left and the control u on the right. Final time $T = 20$ (blue line) and $T = 40$ (red dash-dot line).

First, we point out that the results presented in Table 2 clearly show that the mesh selection based on conditioning allows the solution of the problem using a reduced number of mesh points and vectorial function evaluations. To gain the convergence for the other codes, it can be sufficient, in some cases, to increase the number of points in the initial mesh. To this aim, in Table 3 we show the numerical results obtained for bvp5c using 501 or 1001 initial equidistant points and $T = 2 \times 10^4$. This strategy is advantageous for bvp5c, not yet for bvp4c, that needs an initial mesh of 2501 mesh points to reach the convergence. However, we observe that bvp5c is not able to reach convergence if we use an initial mesh of 2501 mesh points. For the other classes of methods increasing the number of mesh points is not advantageous in terms of computational cost and time execution.

Table 3. Nonlinear Mass spring, initial mesh (IM) with 501, 1001 and 2501 equidistant points and $T = 2 \times 10^4$: final mesh (fM), total number of vectorized function evaluation (NVF) and mixed errors for x, v, u.

	IM	fM	NVF	Error x	Error v	Error u
bvp4c	2501	421	57	9.5×10^{-6}	1.1×10^{-5}	1.0×10^{-5}
bvp5c	501	261	7200	4.2×10^{-6}	4.9×10^{-6}	4.8×10^{-6}
bvp5c	1001	641	13,088	4.1×10^{-6}	4.7×10^{-6}	4.6×10^{-6}
		tol $= 10^{-6}$				
bvp4c	2501	471	71	1.4×10^{-7}	1.4×10^{-7}	1.5×10^{-7}
bvp5c	501	333	7578	3.9×10^{-8}	5.7×10^{-8}	6.1×10^{-8}
bvp5c	1001	512	13,880	3.9×10^{-8}	5.7×10^{-8}	6.1×10^{-8}

To improve the performance of all considered codes, the BVP (3) is reformulated using a variable transformation. Let $\tau = t/T$ with $\tau \in [0,1]$, we solve the following BVP

$$x' = Tv$$
$$v' = -T(x + x^3 + \mu)$$
$$\lambda' = T\left(-x + \mu(1 + 3x^2)\right) \quad (4)$$
$$\mu' = -T(v + \lambda)$$
$$x(0) = 1,\ v(0) = 0,\ x(1) = 0.75,\ v(1) = 0.$$

Now, we set the perturbation parameter $\epsilon = 1/T$, so that we can run for parameters less than 1 the codes acdc and acdcc that use an automatic continuation strategy. For all the other codes, we can adopt a continuation strategy starting with an initial value of ϵ_0 that guarantees the convergence, in our case we use $\epsilon_0 = 1/20$ and we change this value up to reach the required value. To this aim we consider the perturbation parameter changing in the interval $[\epsilon_0, \epsilon]$ among the values $0.5 \times 10^{-j}, j = -2, \ldots, -6$. This means that we discretize the interval with $N_e = 3$ and $N_e = 5$ respectively for $T = 2 \times 10^4$ and $T = 2 \times 10^6$. In Table 4 we show the results obtained applying this successful continuation strategy. All the methods converge for all the values of T using a low computational cost. In this case, the codes based on automatic continuation strategy are very efficient, using acdc/acdcc the users do not need to decide how to change the continuation parameters, even if in some cases the automatic continuation could fail to reach the final desired value.

More information about this problem could be obtained by analyzing the conditioning parameters given in output by the codes twpbvpc_m and tomc, reported in Table 5. As we can see the stiffness parameter σ grows with the width of the interval, and depends on this last, moreover $\kappa_2 > \kappa_1$ shows that the problem could be ill posed, and γ_1 tending to zeros shows the presence of different time scales. The transformation of time interval in $[0,1]$ does not change the stiffness of the problem, but the problem is well posed (see Table 6).

4.2. Completely Hypersensitive Control Problem

This example is a hypersensitive optimal control problem implemented in ICLOCS2, defined as a problem "extremely difficult" to solve using an indirect method [42,44] and given by

$$\begin{cases} \min_{x,u} \int_0^T (x^2 + u^2)\, dt \\ x' = -x^3 + u \\ x(0) = 1,\ x(T) = 1.5. \end{cases} \quad (5)$$

Considered the Hamiltonian $H(x, \lambda, u) = x^2 + u^2 + \lambda(-x^3 + u)$, the first-order necessary conditions for optimality leads to the following boundary value problem (BVP)

$$\begin{aligned} x' &= -x^3 - \frac{\lambda}{2} \\ \lambda' &= -2x + 3\lambda x^2 \\ x(0) &= 1,\ x(T) = 1.5, \end{aligned} \quad (6)$$

where the optimal control is $u^* = -\frac{\lambda}{2}$. We choose $T = 10^4$, $T = 10^6$ and an initial mesh of 11 equidistant points, the solution is plotted in Figure 2. Numerical results shown in Table 7 point out good performance of all the codes except bvp4c and bvp5c, which are not suitable for stiff problems, indeed we underline as they converge to the solution respectively up to $T = 38$ and $T = 29$.

In Table 8 the approximations of the conditioning constants show the dependence of the stiffness on the width of the interval. Moreover, the numerical results underline the necessity of adopting a good mesh selection strategy for computing the solution.

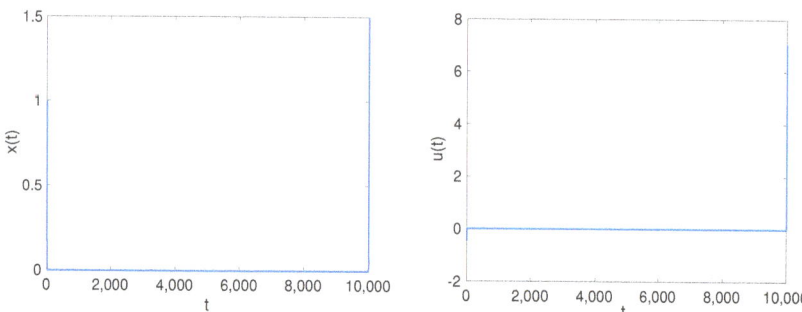

Figure 2. Hypersensitive: solution in time for the mass position x on the left and the control u on the right, final time $T = 10^4$.

Table 4. Nonlinear Mass spring using the variable $\tau = t/T$, initial mesh with starting mesh with 11 equidistant points and continuation strategy on $\epsilon = 1/T$, final mesh (fM), total number of vectorized function evaluation (NVF) and mixed errors for x, v, u.

	T = 20					T = 2 × 10⁴					T = 2 × 10⁶				
	fM	NVF	Error x	Error v	Error u	fM	NVF	Error x	Error v	Error u	fM	NVF	Error x	Error v	Error u
								tol = 10⁻⁴							
bvp4c	78	45	6.3×10^{-6}	9.6×10^{-6}	2.8×10^{-5}	199	178	4.3×10^{-4}	4.5×10^{-4}	5.9×10^{-4}	2093	334	2.5×10^{-3}	3.5×10^{-3}	3.5×10^{-3}
bvp5c	38	220	1.6×10^{-6}	1.1×10^{-6}	9.5×10^{-7}	148	1658	3.3×10^{-6}	7.1×10^{-6}	8.4×10^{-6}	1784	11,712	3.0×10^{-6}	4.5×10^{-6}	4.8×10^{-6}
twpbvp_m	24	56	1.4×10^{-6}	9.2×10^{-7}	1.3×10^{-6}	137	233	6.8×10^{-5}	8.2×10^{-5}	9.2×10^{-5}	163	358	2.6×10^{-6}	2.9×10^{-6}	2.9×10^{-6}
twpbvpc_m	39	81	3.2×10^{-6}	1.9×10^{-6}	2.3×10^{-6}	421	221	3.5×10^{-6}	2.6×10^{-6}	2.0×10^{-6}	141	335	2.5×10^{-5}	4.8×10^{-5}	6.8×10^{-5}
twpbvp_l	28	51	6.3×10^{-6}	9.6×10^{-6}	2.8×10^{-5}	76	346	1.1×10^{-5}	1.4×10^{-5}	1.9×10^{-5}	83	557	1.1×10^{-5}	1.4×10^{-5}	1.9×10^{-5}
twpbvpc_l	28	51	6.3×10^{-6}	9.6×10^{-6}	2.8×10^{-5}	102	237	6.0×10^{-6}	9.4×10^{-6}	1.1×10^{-5}	139	343	6.0×10^{-6}	9.4×10^{-6}	1.1×10^{-5}
tom	156	16	7.3×10^{-7}	1.2×10^{-6}	1.5×10^{-6}	481	49	1.1×10^{-5}	6.7×10^{-6}	1.1×10^{-5}	711	73	7.0×10^{-7}	6.9×10^{-7}	8.1×10^{-7}
tomc	141	16	1.6×10^{-6}	1.1×10^{-6}	2.6×10^{-6}	681	31	1.6×10^{-6}	2.1×10^{-6}	2.4×10^{-6}	1356	49	9.0×10^{-8}	1.7×10^{-7}	2.4×10^{-7}
acdc	25	155	2.8×10^{-5}	2.7×10^{-5}	3.0×10^{-5}	66	485	2.6×10^{-6}	4.0×10^{-6}	6.2×10^{-6}	110	749	2.9×10^{-6}	1.6×10^{-6}	3.3×10^{-6}
acdcc	25	155	2.8×10^{-5}	2.7×10^{-5}	3.0×10^{-5}	106	375	2.3×10^{-5}	1.4×10^{-5}	1.5×10^{-5}	176	501	2.3×10^{-5}	3.0×10^{-5}	3.7×10^{-5}
								tol = 10⁻⁶							
bvp4c	296	57	2.6×10^{-8}	2.5×10^{-8}	2.8×10^{-8}	319	160	1.6×10^{-6}	2.0×10^{-6}	2.6×10^{-6}	374	280	1.5×10^{-5}	1.6×10^{-5}	1.7×10^{-5}
bvp5c	87	464	1.2×10^{-8}	9.3×10^{-9}	4.9×10^{-9}	203	3535	1.4×10^{-8}	1.9×10^{-8}	2.7×10^{-8}	1305	19,524	2.5×10^{-8}	4.0×10^{-8}	4.4×10^{-8}
twpbvp_m	39	83	8.2×10^{-8}	8.7×10^{-8}	9.5×10^{-8}	178	341	7.0×10^{-8}	8.8×10^{-8}	9.2×10^{-8}	497	459	5.4×10^{-9}	7.1×10^{-9}	8.3×10^{-9}
twpbvpc_m	50	83	8.2×10^{-8}	8.7×10^{-8}	9.5×10^{-8}	190	255	7.9×10^{-8}	5.5×10^{-8}	6.4×10^{-9}	423	375	8.1×10^{-9}	8.1×10^{-9}	8.9×10^{-9}
twpbvp_l	50	90	9.8×10^{-9}	1.1×10^{-8}	1.2×10^{-8}	114	284	1.9×10^{-8}	2.5×10^{-8}	1.0×10^{-7}	103	446	1.0×10^{-8}	1.1×10^{-8}	1.7×10^{-8}
twpbvpc_l	61	90	9.8×10^{-9}	1.1×10^{-8}	1.2×10^{-8}	120	246	2.4×10^{-8}	2.2×10^{-8}	3.3×10^{-8}	127	350	2.4×10^{-8}	2.2×10^{-8}	3.3×10^{-8}
tom	161	17	8.0×10^{-7}	1.3×10^{-6}	1.6×10^{-6}	426	62	5.8×10^{-7}	7.6×10^{-7}	9.4×10^{-7}	751	95	2.1×10^{-7}	2.6×10^{-7}	2.9×10^{-7}
tomc	186	19	2.1×10^{-7}	4.1×10^{-7}	4.5×10^{-7}	831	36	2.4×10^{-7}	2.0×10^{-7}	2.5×10^{-7}	826	56	2.2×10^{-7}	2.3×10^{-7}	2.6×10^{-7}
acdc	50	158	1.4×10^{-8}	1.8×10^{-8}	2.1×10^{-8}	95	382	1.0×10^{-8}	1.8×10^{-8}	2.0×10^{-8}	89	581	2.7×10^{-8}	3.8×10^{-8}	4.4×10^{-8}
acdcc	50	158	1.4×10^{-8}	1.8×10^{-8}	2.1×10^{-8}	183	371	1.8×10^{-7}	2.0×10^{-7}	2.1×10^{-7}	184	513	1.8×10^{-7}	2.0×10^{-7}	2.1×10^{-7}

Table 5. Nonlinear Mass spring: conditioning parameters computed using $tol = 10^{-6}$ and initial mesh with 11 equidistant points.

	σ	κ	κ_1	κ_2	γ_1
		$T = 20$			
twpbvpc_m	1.90×10^1	1.33×10^1	4.72×10^0	8.57×10^0	2.16×10^{-1}
tomc	2.04×10^1	1.33×10^1	4.79×10^0	8.61×10^0	5.28×10^{-1}
		$T = 2 \times 10^4$			
twpbvpc_m	1.97×10^4	1.34×10^1	4.73×10^0	8.65×10^0	5.09×10^{-4}
tomc	2.08×10^4	1.66×10^1	5.42×10^0	1.12×10^1	2.11×10^{-4}
		$T = 2 \times 10^6$			
twpbvpc_m	1.90×10^6	1.34×10^1	4.75×10^0	8.61×10^0	5.28×10^{-6}
tomc	2.03×10^6	1.66×10^1	5.45×10^0	1.11×10^1	2.17×10^{-6}

Table 6. Nonlinear Mass spring using the variable $\tau = t/T$: conditioning parameters computed using $tol = 10^{-6}$ and initial mesh with 11 equidistant points.

	σ	κ	κ_1	κ_2	γ_1
		$T = 20$			
twpbvpc_m	1.90×10^1	5.15×10^0	4.72×10^0	4.31×10^{-1}	5.28×10^{-1}
tomc	2.06×10^1	5.18×10^0	4.75×10^0	4.30×10^{-1}	2.12×10^{-1}
		$T = 2 \times 10^4$			
twpbvpc_m	1.90×10^4	4.73×10^0	4.73×10^0	4.30×10^{-4}	5.26×10^{-4}
tomc	2.08×10^4	4.75×10^0	4.75×10^0	4.31×10^{-4}	2.10×10^{-4}
		$T = 2 \times 10^6$			
twpbvpc_m	1.90×10^6	4.75×10^0	4.75×10^0	4.31×10^{-6}	5.28×10^{-6}
tomc	2.09×10^6	4.75×10^0	4.75×10^0	4.31×10^{-6}	2.09×10^{-6}

Table 7. Hypersensitive problem solved with an initial mesh with 11 equidistant points: final mesh (fM), total number of vectorized function evaluation (NVF) and mixed errors for x, v, u.

	fM	NVF	Error x	Error v	Error u	fM	NVF	Error x	Error v	Error u
twpbvp_m	117	239	1.5×10^{-6}	3.1×10^{-6}	1.5×10^{-6}	1821	394	1.9×10^{-5}	3.8×10^{-5}	1.9×10^{-5}
twpbvpc_m	140	237	1.7×10^{-6}	3.4×10^{-6}	1.7×10^{-6}	650	456	3.3×10^{-5}	6.7×10^{-5}	5.0×10^{-5}
twpbvp_l	105	224	7.0×10^{-6}	2.7×10^{-5}	1.7×10^{-5}	*	*	*	*	*
twpbvpc_l	91	286	6.6×10^{-9}	1.3×10^{-8}	6.6×10^{-9}	1176	425	8.2×10^{-9}	3.0×10^{-8}	1.9×10^{-8}
tom	691	33	2.8×10^{-9}	5.5×10^{-9}	2.8×10^{-9}	636	169	2.5×10^{-6}	2.3×10^{-5}	2.1×10^{-5}
tomc	681	46	5.7×10^{-7}	1.1×10^{-6}	5.7×10^{-7}	1941	134	4.0×10^{-9}	8.0×10^{-9}	4.0×10^{-9}
					$tol = 10^{-6}$					
twpbvp_m	357	245	9.8×10^{-9}	2.0×10^{-8}	1.3×10^{-8}	1859	392	2.6×10^{-8}	5.3×10^{-8}	2.6×10^{-8}
twpbvpc_m	265	239	3.2×10^{-7}	6.4×10^{-7}	3.2×10^{-7}	536	475	4.4×10^{-8}	8.8×10^{-8}	4.4×10^{-8}
twpbvp_l	94	248	5.3×10^{-8}	1.1×10^{-7}	5.3×10^{-8}	*	*	*	*	*
twpbvpc_l	91	286	6.6×10^{-9}	1.3×10^{-8}	6.6×10^{-9}	1176	425	8.2×10^{-9}	3.0×10^{-8}	1.9×10^{-8}
tom	691	33	2.8×10^{-9}	5.5×10^{-9}	2.8×10^{-9}	691	172	8.8×10^{-8}	2.5×10^{-7}	2.3×10^{-7}
tomc	681	46	5.7×10^{-7}	1.1×10^{-6}	5.7×10^{-7}	1941	134	4.0×10^{-9}	8.0×10^{-9}	4.0×10^{-9}

Table 8. Hypersensitive problem: conditioning parameters computed using $tol = 10^{-6}$ and initial mesh with 11 equidistant points.

	σ	κ	κ_1	κ_2	γ_1
			$T = 10$		
twpbvpc_m	5.94×10^1	3.09×10^1	2.67×10^1	4.23×10^0	5.51×10^{-1}
twpbvpc_l	5.95×10^1	3.09×10^1	2.67×10^1	4.23×10^0	5.49×10^{-1}
tomc	6.83×10^1	3.09×10^1	2.67×10^1	4.23×10^0	3.90×10^{-1}
			$T = 10^4$		
twpbvpc_m	6.34×10^4	3.09×10^1	2.66×10^1	4.23×10^0	5.24×10^{-4}
twpbvpc_l	5.80×10^4	3.09×10^1	2.67×10^1	4.23×10^0	5.61×10^{-4}
tomc	6.74×10^4	3.09×10^1	2.66×10^1	4.23×10^0	3.96×10^{-4}
			$T = 10^6$		
twpbvpc_m	6.44×10^6	3.09×10^1	2.66×10^1	4.23×10^0	5.11×10^{-6}
twpbvpc_l	5.97×10^6	3.09×10^1	2.67×10^1	4.23×10^0	5.54×10^{-6}
tomc	6.92×10^6	4.70×10^1	9.15×10^0	3.78×10^1	3.85×10^{-6}

For the purpose of improving the performance and overcoming some drawbacks, we propose, as already done for the test problem in Section 4.1, to use the transformation of the variable $\tau = t/T$, such that the BVP (6) can be reformulated for $\tau \in [0,1]$ as

$$x' = T\left(-x^3 - \frac{\lambda}{2}\right)$$
$$\lambda' = T\left(-2x + 3\lambda x^2\right) \qquad (7)$$
$$x(0) = 1, \; x(1) = 1.5.$$

The advantage of this formulation is that considering $\epsilon = 1/T$ as a perturbation parameter, we can apply the continuation strategy on that parameter. In Table 9 we report the results using as starting value $\epsilon_0 = 1/10$ and changing the continuation parameters in the interval $[\epsilon_0, \epsilon]$ among the value of the set $10^{-2}, 10^{-3}, 10^{-4}$. We remember that acdc and acdcc, using an automatic continuation strategy, needs only to insert the desired value of ϵ and uses as ϵ_0 the default value 0.5. The numerical tests and the conditioning parameters in Tables 8 and 10 clearly show that for this class of problems, if we cannot use a continuation of parameters, the codes able to give a solution are the ones suited for stiff problems that work still better if also the mesh selection is appropriate for this class of problems.

Table 9. Hypersensitive problem using the variable $\tau = t/T$, initial mesh with 11 equidistant points and continuation strategy on T: final mesh (fM), total number of vectorized function evaluation (NVF) and mixed errors for x, v, u.

					$T = 10^4$				
			$tol = 10^{-4}$					$tol = 10^{-6}$	
	fM	NVF	Error x	Error v	Error u	fM	NVF	Error x	Error v	Error u
bvp4c	107	198	1.1×10^{-4}	4.5×10^{-4}	3.8×10^{-4}	242	170	1.1×10^{-6}	4.6×10^{-6}	3.6×10^{-6}
bvp5c	70	1277	8.5×10^{-7}	1.7×10^{-6}	8.5×10^{-7}	132	2874	4.9×10^{-9}	9.8×10^{-9}	4.9×10^{-9}
twpbvp_m	59	266	2.2×10^{-5}	4.4×10^{-5}	2.2×10^{-5}	124	311	4.9×10^{-6}	8.1×10^{-6}	4.9×10^{-6}
twpbvpc_m	101	210	9.1×10^{-5}	1.8×10^{-4}	9.1×10^{-5}	190	226	9.5×10^{-8}	1.9×10^{-7}	9.5×10^{-8}
twpbvp_l	45	389	3.5×10^{-6}	1.9×10^{-5}	1.8×10^{-5}	63	306	4.9×10^{-8}	1.9×10^{-7}	1.8×10^{-7}
twpbvpc_l	76	229	1.3×10^{-5}	2.6×10^{-5}	1.3×10^{-5}	79	234	4.7×10^{-8}	9.4×10^{-8}	4.7×10^{-8}
tom	571	56	7.9×10^{-7}	1.8×10^{-7}	1.2×10^{-7}	746	59	1.1×10^{-8}	2.6×10^{-8}	2.1×10^{-8}
tomc	951	46	4.9×10^{-9}	7.2×10^{-9}	5.8×10^{-9}	1231	52	1.2×10^{-8}	3.7×10^{-9}	2.3×10^{-9}
acdc	56	491	4.0×10^{-6}	1.5×10^{-5}	1.4×10^{-5}	60	425	5.7×10^{-8}	1.9×10^{-7}	1.7×10^{-7}
acdcc	130	672	2.1×10^{-5}	5.6×10^{-5}	3.6×10^{-5}	144	478	5.2×10^{-8}	1.8×10^{-7}	1.2×10^{-7}

Table 10. Hypersensitive problem using the variable $\tau = t/T$: conditioning parameters computed using $tol = 10^{-6}$ and initial mesh with 11 equidistant points.

	σ	κ	κ_1	κ_2	γ_1
			$T = 10$		
twpbvpc_m	5.94×10^1	2.71×10^1	2.67×10^1	4.23×10^{-1}	5.51×10^{-1}
twpbvpc_l	5.95×10^1	2.71×10^1	2.67×10^1	4.23×10^{-1}	5.49×10^{-1}
tomc	6.86×10^1	2.71×10^1	2.66×10^1	4.23×10^{-1}	3.88×10^{-1}
			$T = 10^4$		
twpbvpc_m	6.34×10^4	2.66×10^1	2.66×10^1	4.23×10^{-4}	5.24×10^{-4}
twpbvpc_l	5.80×10^4	2.67×10^1	2.67×10^1	4.23×10^{-4}	5.61×10^{-4}
tomc	6.74×10^4	2.66×10^1	2.66×10^1	4.23×10^{-4}	3.95×10^{-4}
			$T = 10^6$		
twpbvpc_m	6.52×10^6	2.66×10^1	2.66×10^1	4.23×10^{-6}	5.06×10^{-6}
twpbvpc_l	5.72×10^6	2.67×10^1	2.67×10^1	4.23×10^{-6}	5.72×10^{-6}
tomc	6.87×10^6	2.66×10^1	2.66×10^1	4.23×10^{-6}	3.88×10^{-6}

5. Bang-Bang Optimal Control Problem

The bang-bang optimal control problem [45] is among the more challenging ones. It arises from a model in which a point unit mass m subjects to a limited force in one-dimensional space, i.e., $mx''(t) = u(t)$ and $u(t) \leq 1$. The main feature of optimal control problem of moving the mass from $x = 0$ to the maximum distance x in one second can be formulated as follows

$$\min -x(1) = \int_0^1 (-v)dt,$$
$$x' = v,$$
$$v' = u, \quad t \in [0,1], \tag{8}$$
$$x(0) = v(0) = v(1) = 0,$$
$$|u| \leq 1,$$

The associated Hamiltonian function is defined as

$$H(x, v, \lambda, \mu, u) = -v + \lambda v + \mu u$$

and the optimal control is given by

$$u^* = \arg\min_{|u|\leq 1} H(x, v, \lambda, \mu, u) = -sign(\mu).$$

Now, by applying the indirect method the solution of the optimal control Problem (8) is equivalent to solve the following BVP problem

$$x' = v,$$
$$v' = u,$$
$$\lambda' = 0, \tag{9}$$
$$\mu' = 1 - \lambda,$$
$$x(0) = v(0) = v(1) = \lambda(1) = 0.$$

We observe that the optimal control is defined as

$$u(t) = -\text{sign}(\mu) = \begin{cases} 1 & \mu < 0, \\ -1 & \mu > 0, \\ \text{any value in [-1,1]} & \mu = 0, \end{cases}$$

and the exact solution is given by

$$x(t) = \begin{cases} \dfrac{t^2}{2} & t < 1/2 \\ t - \dfrac{t^2}{2} - \dfrac{1}{4} & t > 1/2 \end{cases}, \quad v(t) = \begin{cases} t & t < 1/2 \\ 1 - t & t > 1/2 \end{cases},$$

$$u(t) = \begin{cases} 1 & t < 1/2, \\ -1 & t > 1/2, \end{cases} \quad \lambda(t) = 0, \quad \mu(t) = t - \dfrac{1}{2}.$$

The discontinuity of the switching function is overcome by a smoothing technique that can be executed by different strategies. We choose two of them in particular. The first strategy, given a small parameter ϵ, consists of using the approximation

$$\text{sign}(\mu) \approx \dfrac{2}{\pi} \arctan\left(\dfrac{\mu\pi}{2\epsilon}\right).$$

The exact bang-bang solution is better approximated when ϵ becomes smaller; however, this for value around smaller than 10^{-4} can give ill-conditioning problems. Table 11 contains all the results obtained using the Matlab codes, the solution is plotted in Figure 3. Only bvp5c fails, and for getting the solution is necessary to use the continuation strategy. To this regard we consider as initial perturbation parameter $\epsilon_0 = 1$ and then we change it choosing $N_\epsilon = 10$ logarithmically equispaced points between 1 and the value required ϵ. When $tol = 10^{-4}$ bvp5c converges using 19 points for both ϵ equal to 10^{-3} and 10^{-6}, instead when $tol = 10^{-4}$ bvp5c gets the solution with 36 and 28 points respectively for $\epsilon = 10^{-3}$ and $\epsilon = 10^{-6}$.

Table 11. Bang-Bang optimal control Problem (9): final mesh (fM), total number of vectorized function evaluation (NVF) and mixed errors for x, v, u.

	\multicolumn{9}{c}{$tol = 10^{-4}$}								
			$\epsilon = 10^{-3}$					$\epsilon = 10^{-6}$	
	fM	NVF	Error x	Error v	Error u	fM	NVF	Error x	Error v	Error u
bvp4c	25	51	3.2×10^{-4}	1.8×10^{-3}	0	27	97	3.2×10^{-7}	4.0×10^{-6}	0
twpbvp_m	16	11	3.0×10^{-4}	7.5×10^{-4}	0	16	11	2.1×10^{-5}	7.5×10^{-7}	0
twpbvp_l	16	13	2.5×10^{-4}	7.5×10^{-4}	0	16	13	7.6×10^{-5}	7.5×10^{-7}	0
tom	111	10	3.2×10^{-4}	1.0×10^{-3}	0	31	16	1.8×10^{-4}	8.8×10^{-4}	0
tomc	121	10	3.2×10^{-4}	1.0×10^{-3}	0	31	28	1.8×10^{-4}	8.8×10^{-4}	0
acdc	9	157	3.2×10^{-4}	8.7×10^{-4}	0	9	221	3.2×10^{-7}	1.5×10^{-6}	0
	\multicolumn{9}{c}{$tol = 10^{-6}$}								
bvp4c	79	57	3.2×10^{-4}	2.0×10^{-3}	0	47	93	3.2×10^{-7}	4.0×10^{-6}	0
twpbvp_m	10	32	3.3×10^{-4}	1.0×10^{-3}	0	10	32	5.1×10^{-6}	1.0×10^{-6}	0
twpbvp_l	17	66	3.2×10^{-4}	1.3×10^{-3}	0	15	130	3.2×10^{-7}	2.1×10^{-6}	0
tom	231	19	3.2×10^{-4}	1.1×10^{-3}	0	281	32	3.3×10^{-7}	1.1×10^{-6}	0
tomc	201	19	3.2×10^{-4}	1.1×10^{-3}	0	231	40	3.3×10^{-7}	1.1×10^{-6}	0
acdc	20	170	3.2×10^{-4}	1.3×10^{-3}	0	17	242	3.2×10^{-7}	2.2×10^{-6}	0

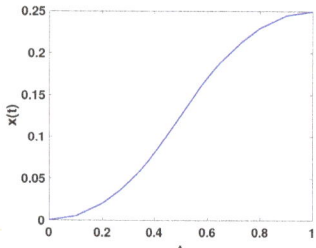 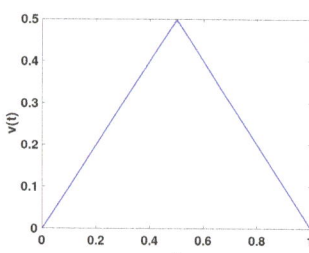 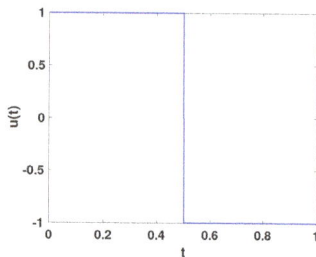

Figure 3. Bang-Bang, $\epsilon = 10^{-3}$: solution in time for the mass position x on the (**left**), for the velocity in the (**center**) and the control u on the (**right**).

For the second smoothing technique we can add a barrier or a penalty function. In this regard, we consider a piecewise quadratic penalty function defined as in [45]

$$P(u; \epsilon, \sigma) = \frac{\epsilon}{2} u^2 + \frac{1}{\sigma^2} \begin{cases} (|u| - 1 + \sigma)^2 & |u| > 1 - \sigma, \\ 0 & \text{otherwise} \end{cases}$$

where the parameter σ gives the distance from the border where the penalty changes fast. Consequently, the Problem (8) is reformulated without inequality constraint as follows

$$\min \int_0^1 P(u; \epsilon, \sigma) - v \, dt,$$
$$x' = v,$$
$$v' = u, \quad t \in [0, 1], \quad (10)$$
$$x(0) = v(0) = v(1) = 0.$$

The optimal control u, obtained as a solution of the equation

$$P_u(u; \epsilon, \sigma) + \mu = 0,$$

is equal to

$$u = \begin{cases} \dfrac{2 - 2\sigma - \sigma^2 \mu}{\epsilon \sigma^2 + 2} & u \geq 1 - \sigma \\ \dfrac{-2 + 2\sigma - \sigma^2 \mu}{\epsilon \sigma^2 + 2} & u < \sigma - 1 \\ 0 & \text{otherwise.} \end{cases}$$

In Table 12 we show the numerical results obtained for $\sigma = 10^{-4}$ and $\epsilon = 10^{-4}, 10^{-6}$, starting with an initial mesh of 16 equidistant points and a null initial solution. It is clear that all the codes have a good performance, we do not report the results for bvp4c and bvp5c because they fail. To overcome this drawback in Table 13 we consider the continuation strategy, this means that the codes bvp4c and bvp5c are run for different values of ϵ starting from $\epsilon_0 = 10$ up to the desired value ϵ. In particular, we choose $N_\epsilon = 10$ values logarithmically equispaced.

Table 12. Bang-Bang optimal control problem-solving (8) using a piecewise quadratic penalty function with $\sigma = 10^{-4}$: final mesh (fM), total number of vectorized function evaluation (NVF) and mixed errors for x, v, u.

			$\epsilon = 10^{-3}$					$\epsilon = 10^{-6}$		
	fM	NVF	Error x	Error v	Error u	fM	NVF	Error x	Error v	Error u
					$tol = 10^{-4}$					
twpbvp_m	16	11	9.8×10^{-6}	3.2×10^{-5}	5.0×10^{-5}	16	11	9.8×10^{-6}	3.2×10^{-5}	5.0×10^{-5}
twpbvp_l	16	13	6.4×10^{-5}	3.2×10^{-5}	5.0×10^{-5}	16	13	6.4×10^{-5}	3.2×10^{-5}	5.0×10^{-5}
tom	111	10	2.2×10^{-5}	6.3×10^{-5}	5.0×10^{-5}	31	27	1.9×10^{-4}	8.5×10^{-4}	5.0×10^{-5}
tomc	126	9	2.1×10^{-5}	6.6×10^{-5}	5.0×10^{-5}	31	20	1.9×10^{-4}	8.5×10^{-4}	5.0×10^{-5}
					$tol = 10^{-6}$					
twpbvp_m	8	32	2.4×10^{-5}	3.3×10^{-5}	5.0×10^{-5}	8	32	2.4×10^{-5}	3.3×10^{-5}	5.0×10^{-5}
twpbvp_l	9	93	2.0×10^{-5}	3.3×10^{-5}	5.0×10^{-5}	8	130	2.0×10^{-5}	3.7×10^{-5}	5.0×10^{-5}
tom	231	19	2.0×10^{-5}	3.4×10^{-5}	5.0×10^{-5}	381	51	2.0×10^{-5}	3.3×10^{-5}	5.0×10^{-5}
tomc	261	20	2.0×10^{-5}	3.3×10^{-5}	5.0×10^{-5}	241	56	2.0×10^{-5}	3.3×10^{-5}	5.0×10^{-5}

Table 13. Bang-Bang optimal control problem-solving (8) using a piecewise quadratic penalty function with $\sigma = 10^{-4}$ and the continuation strategy: final mesh (fM), total number of vectorized function evaluation (NVF) and mixed errors for x, v, u.

				$\epsilon = 10^{-3}$						$\epsilon = 10^{-6}$		
	N_ϵ	fM	NVF	Error x	Error v	Error u	N_ϵ	fM	NVF	Error x	Error v	Error u
						$tol = 10^{-4}$						
bvp4c	10	12	1899	2.0×10^{-5}	3.3×10^{-5}	5.0×10^{-5}	5	13	1214	2.0×10^{-5}	2.0×10^{-5}	5.0×10^{-5}
bvp5c	10	9	1551	2.0×10^{-5}	3.3×10^{-5}	5.0×10^{-5}	10	13	1442	2.0×10^{-5}	3.3×10^{-5}	5.0×10^{-5}
acdc		4	326	1.5×10^{-5}	3.3×10^{-5}	5.0×10^{-5}		4	326	1.5×10^{-5}	3.3×10^{-5}	5.0×10^{-5}
						$tol = 10^{-6}$						
bvp4c	10	16	3168	2.0×10^{-5}	3.3×10^{-4}	5.0×10^{-5}	10	19	3421	2.0×10^{-5}	2.0×10^{-5}	5.0×10^{-5}
bvp5c	10	13	3235	2.0×10^{-5}	3.3×10^{-5}	5.0×10^{-5}	100	14	21747	2.0×10^{-5}	3.3×10^{-5}	5.0×10^{-5}
acdc		9	380	2.0×10^{-5}	6.7×10^{-5}	5.0×10^{-5}		9	380	2.0×10^{-5}	3.3×10^{-5}	5.0×10^{-5}

We also report the results of acdc and acdcc that use an automatic continuation strategy. The results point out the suitability and efficiency of the strategy in solving this kind of problems, also for bvp4c and bvp5c when the nonlinear solution is approximated using a continuation strategy. The conditioning parameters reported in Tables 14 and 15 show that the problem is not stiff since σ is of moderate size, indeed the main difficulty is caused by the convergence of the nonlinear discretization schemes. In this regard we highlight as the results of the codes twpbvpc_m and twpbvpc_l are the same of those gained by the codes twpbvp_m and twpbvp_l, confirming the non-necessity of these codes to use a mesh selection strategy based on conditioning for this non-stiff problem.

Table 14. Bang-Bang optimal control problem: conditioning parameters computed using $tol = 10^{-6}$.

	σ	κ	κ_1	κ_2	γ_1
		$\epsilon = 10^{-3}$			
twpbvpc_m	1.8	3.3	2.0	1.3	1.6
twpbvpc_l	1.4	3.3	2.0	1.3	1.6
tomc	1.5	3.3	2.0	1.2	1.0
		$\epsilon = 10^{-6}$			
twpbvpc_m	2.0	3.2	2.0	1.2	1.6
twpbvp_l	2.0	3.3	2.0	1.3	1.7
tomc	2.1	3.3	2.0	1.2	1.0

Table 15. Bang-Bang optimal control problem with penalty: conditioning parameters computed using $tol = 10^{-6}$.

	σ	κ	κ_1	κ_2	γ_1
		$\epsilon = 10^{-3}$			
twpbvpc_m	1.8	3.2	2.0	1.2	1.6
twpbvpc_l	1.4	3.3	2.0	1.3	1.7
tomc	1.4	3.3	2.0	1.2	1.0
		$\epsilon = 10^{-6}$			
twpbvpc_m	2.0	3.2	2.0	1.2	1.6
twpbvpc_l	2.0	3.3	2.0	1.3	1.7
tomc	1.3	3.3	2.0	1.2	1.0

6. Longitudinal Dynamics of a Vehicle

We consider an example of nonlinear optimal control problem derived from a model of the longitudinal dynamics of a vehicle with the aerodynamic down-force [2]. In particular, a vehicle, supposed to be a point mass, is moved in a fixed time T from an initial zero velocity to a final zero velocity

$$\min\{x(0) - x(T)\} = \min\left(-\int_0^T v\,dt\right)$$

$$x' = v,$$
$$v' = u - k_0 - k_1 v - k_2 v^2, \qquad t \in [0, T], \tag{11}$$
$$x(0) = v(0) = v(T) = 0,$$
$$|u| \leq g + k_3 v^2.$$

The Hamiltonian function associated with this problem is

$$H(x, v, \lambda, \mu, u) = -v + \lambda v + \mu\left(u - k_0 - k_1 v - k_2 v^2\right)$$

and the optimal control is given by

$$u^* = \underset{|u| \leq g + k_3 v^2}{\arg\min}\, H(x, v, \lambda, \mu, u) = -(g + k_3 v^2)\operatorname{sign}(\mu).$$

Now, applying the indirect method the global optimal control problem is reduced to the boundary value problem

$$x' = v,$$
$$v' = u - k_0 - k_1 v - k_2 v^2,$$
$$\lambda' = 0, \qquad (12)$$
$$\mu' = 1 - \lambda + \mu(k_1 + 2vk_2),$$
$$x(0) = v(0) = v(T) = \lambda(T) = 0.$$

We observe that the optimal control problem has a theoretical solution given by

$$u = -sign(\mu)(g + k_3 v^2)$$

that can be approximated using a barrier function defined as

$$u = -\frac{2}{\pi}(g + k_3 v^2) \arctan\left(\frac{2\mu}{\pi \epsilon}\right).$$

Let $g_+ = g + k_0$ and $g_- = g - k_0$, if $t_s = \frac{1}{k_1} \ln \frac{g_- + g_+ e^{k_1 T}}{2g}$ is the switching time, then the solution for the optimal control is defined as

$$u(t) = \begin{cases} 1 & t \leq t_s, \\ -1 & t > t_s. \end{cases}$$

Moreover, the exact solution for the space and the velocity is expressed by

$$x(t) = \begin{cases} k_1^{-2} g_- \left(k_1 t + e^{-k_1 t} - 1\right) & t \leq t_s, \\ k_1^{-2}\left(g_+ + e^{-k_1 t}\left(g_- - 2g e^{k_1 t_s}\right) + k_1 (2g t_s - t g_+)\right) & t > t_s, \end{cases}$$

and

$$v(t) = \begin{cases} k_1^{-1} g_- \left(1 - e^{-k_1 t}\right) & t \leq t_s, \\ k_1^{-1} g_+ \left(e^{k_1(T-t)} - 1\right) & t > t_s, \end{cases}$$

while the multipliers assume the form

$$\lambda(t) = 0, \quad \mu(t) = \frac{1}{k_1}\left(\frac{2g e^{k_1(t-T)}}{g_- e^{-k_1 t} + g_+} - 1\right).$$

In Table 16 are shown all the numerical results obtained using all the Matlab codes considered starting with an initial mesh of 11 equispaced points and an initial approximation with null elements, the solution is plotted in Figure 4. For this problem only the codes of the bvptwp package are able to give a solution for $\epsilon = 10^{-6}$, so for the other codes we have used a continuation strategy with a starting value $\epsilon_0 = 10^{-3}$ and $N_\epsilon = 10$ logarithmic equispaced intermediate points. In Table 16 all the results obtained are shown in order that the symbol c in bracket labels those computed using the continuation strategy. Moreover, the results emphasize that not always the automatic continuation is advantageous and cheaper from a computational cost of view, since it is evident that the total number of vectorial functions evaluation is much greater for acdc than for twpbvp_m and twpbvp_l. Remember that they use the same numerical scheme. The conditioning parameters in Table 17 are all moderate size, hence the problem is not stiff.

Table 16. Longitudinal dynamics of a vehicle $T = 10$, $g = 9.81$, $k_0 = 0.02\,g$, $k_1 = 10^{-5}g$, $k_2 = 0$, $k_3 = 0$: final mesh (fM), total number of vectorized function evaluation (NVF) and mixed errors for x, v, u.

			$tol = 10^{-4}$							
			$\epsilon = 10^{-3}$					$\epsilon = 10^{-6}$		
	fM	NVF	Error x	Error v	Error u	fM	NVF	Error x	Error v	Error u
bvp4c	38	125	4.4×10^{-4}	1.8×10^{-3}	0	32(c)	242	4.4×10^{-7}	3.1×10^{-6}	0
bvp5c	18	270	4.4×10^{-4}	1.7×10^{-3}	0	18(c)	662	4.4×10^{-7}	2.1×10^{-6}	0
twpbvp_m	38	54	4.4×10^{-4}	1.6×10^{-3}	0	13	146	4.0×10^{-7}	5.2×10^{-4}	0
twpbvp_l	38	57	4.4×10^{-4}	2.0×10^{-3}	0	28	134	4.0×10^{-7}	7.4×10^{-4}	0
tom	176	23	4.4×10^{-4}	1.6×10^{-3}	0	176(c)	50	4.7×10^{-7}	4.0×10^{-5}	0
tomc	131	20	4.4×10^{-4}	1.6×10^{-3}	0	131(c)	48	1.1×10^{-6}	2.3×10^{-4}	0
acdc	15	320	4.4×10^{-4}	1.1×10^{-3}	0	8	550	2.0×10^{-6}	1.1×10^{-6}	0

Table 17. Longitudinal dynamics of a vehicle $T = 10$, $g = 9.81$, $k_0 = 0.02\,g$, $k_1 = 10^{-5}g$, $k_2 = 0$, $k_3 = 0$: conditioning parameters computed using $tol = 10^{-6}$.

	σ	κ	κ_1	κ_2	γ_1
		$\epsilon = 10^{-3}$			
twpbvpc_m	3.04×10^0	4.88×10^1	1.11×10^1	4.36×10^1	7.22×10^0
twpbvpc_l	3.11×10^0	4.89×10^1	1.11×10^1	4.36×10^1	7.15×10^0
tomc	3.78×10^0	2.94×10^2	7.43×10^1	2.20×10^2	4.02×10^0
		$\epsilon = 10^{-6}$			
twpbvpc_m	3.81×10^0	4.84×10^1	1.10×10^1	4.33×10^1	6.47×10^0
twpbvpc_l	3.83×10^0	4.84×10^1	1.10×10^1	4.33×10^1	6.47×10^0
tomc	3.86×10^0	2.99×10^2	7.48×10^1	2.24×10^2	4.02×10^0

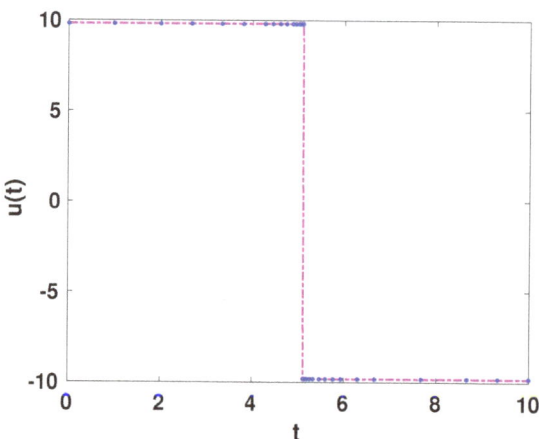

Figure 4. Longitudinal dynamics of a vehicle, $\epsilon = 10^{-3}$, $T = 10$, $g = 9.81$, $k_0 = 0.02\,g$, $k_1 = 10^{-5}g$, $k_2 = 0$, $k_3 = 0$: theoretical (dash-dot line) and numerical (dot line) solution in time for the control u.

7. Gottard Rocket

Now, we consider an example of optimal control problem with a singular arc [46]. A rocket of mass m lifts off vertically at time $t = 0$ with (normalized) altitude $h(0) = 1$ and velocity $v(0) = 0$. Known the initial mass, the fuel mass and the drag characteristics of the rocket, the aim is to choose the thrust $u(t)$ and the final time T to maximize the altitude $h(T)$ at the final time T. The optimal control problem is given by

$$\min_{T,v} \int_0^T (-v)dt,$$
$$h' = v,$$
$$v' = \frac{u - D(h,v)}{m} - g(h), \tag{13}$$
$$m' = -\frac{u}{c},$$
$$0 \leq u \leq u_{max},$$
$$h(0) = 1, \ v(0) = 0, \ m(0) = 1, \ m(T) = 0.6.$$

Given the constants D_c and h_c, the aerodynamic drag is defined by

$$D(h,v) = D_c v^2 e^{-h_c \left(\frac{h - h(0)}{h(0)} \right)}.$$

Moreover, if g_0 is the gravitational force at the earth's surface, then the gravitational force is given by

$$g(h) = g_0 \left(\frac{h(0)}{h} \right)^2.$$

The equation is scaled choosing the model parameters $m(0)$, $h(0)$ and g_0, which allows management of dimension-free equations. As in [46], we consider

$$u_{max} = 3.5 g_0 m(0), \quad D_c = \frac{1}{2} v_c \frac{m(0)}{g_0}, \quad c = \frac{1}{2}(g_0 h(0))^{1/2},$$

where $g_0 = 1$, $h_c = 500$, $m_c = 0.6$ and $v_c = 620$.

Since the problem (13) has a free final time, we fix the time interval using the variable transformation $t(\tau) := \tau T$, with $\tau \in [0,1]$. A new state variable T satisfying the differential constrain $\dot{T} = 0$ is added to the problem and a penalty function $P(u; \epsilon, \bar{\sigma})$ is used as smoothing technique, so that the problem can be reformulated as follows

$$\min_{T,v,u} \int_0^1 \left(-Tv + TP(u; \epsilon, \bar{\sigma}) \right) d\tau,$$
$$h' = Tv,$$
$$v' = \frac{T}{m}\left(u - \frac{1}{2} v_c v^2 e^{h_c(1-h)} \right) - \frac{T}{h^2}, \tag{14}$$
$$m' = -T\frac{u}{c},$$
$$T' = 0,$$
$$h(0) = 1, \ v(0) = 0, \ m(0) = 1, \ m(1) = 0.6.$$

As in Section 5, $P(u; \epsilon, \bar{\sigma})$ is a piecewise quadratic penalty function defined as

$$P(u; \epsilon, \bar{\sigma}) = \frac{\epsilon}{2}\left(u - \frac{u_{max}}{2} \right)^2 + \frac{1}{\bar{\sigma}^2} \begin{cases} (u - u_{max} + \bar{\sigma})^2 & u > u_{max} - \bar{\sigma} \\ (\bar{\sigma} - u)^2 & u < \bar{\sigma} \\ 0 & \text{otherwise}. \end{cases}$$

Now, the Hamiltonian formulation of the problem (14) gives as a result the following BVP

$$h' = Tv,$$
$$v' = \frac{T}{m}\left(u - \frac{1}{2}v_c v^2 e^{h_c(1-h)}\right) - \frac{T}{h^2},$$
$$m' = -T\frac{u}{c},$$
$$T' = 0,$$
$$\lambda_1' = -T\lambda_2\left(\frac{1}{2m}h_c v_c v^2 e^{h_c(1-h)} + \frac{2}{h^3}\right), \quad (15)$$
$$\lambda_2' = -T\left(\lambda_1 - \lambda_2 \frac{v_c v e^{h_c(1-h)}}{m} - 1\right),$$
$$\lambda_3' = \frac{T}{m^2}\lambda_2\left(u - \frac{1}{2}v_c v^2 e^{h_c(1-h)}\right),$$
$$\lambda_4' = v - P(u;\epsilon,\bar{\sigma}) - \lambda_1 v - \lambda_2\left(\frac{u - \frac{1}{2}v_c v^2 e^{h_c(1-h)}}{m} - \frac{1}{h^2}\right) + \lambda_3\frac{u}{c},$$
$$h(0) = 1, \ v(0) = 0, \ m(0) = 1, \ m(1) = 0.6,$$
$$\lambda_1(1) = 0, \ \lambda_2(1) = 0, \ \lambda_4(0) = 0, \ \lambda_4(1) = 0,$$

where the thrust u, computed by solving the equation

$$P_u(u;\epsilon,\bar{\sigma}) + \frac{\lambda_2}{m} - \frac{\lambda_3}{c} = 0,$$

is equivalent to

$$u = \begin{cases} \frac{1}{\epsilon\bar{\sigma}^2 + 2}\left(\epsilon\bar{\sigma}^2\frac{u_{max}}{2} + 2u_{max} - 2\bar{\sigma} + \bar{\sigma}^2\left(\frac{\lambda_2}{m} - \frac{\lambda_3}{c}\right)\right) & u > u_{max} - \bar{\sigma} \\ \frac{\bar{\sigma}}{\epsilon\bar{\sigma}^2 + 2}\left(\epsilon\bar{\sigma}\frac{u_{max}}{2} + 2 - \bar{\sigma}\left(\frac{\lambda_2}{m} - \frac{\lambda_3}{c}\right)\right) & u < \bar{\sigma} \\ \frac{1}{\epsilon}\left(\epsilon\frac{u_{max}}{2} - \frac{\lambda_2}{m} + \frac{\lambda_3}{c}\right) & \text{otherwise.} \end{cases}$$

Since the problem is highly nonlinear, it is chosen as starting approximation of the solution $h = \lambda_1 = \lambda_2 = \lambda_3 = 1, \lambda_4 = 0, v(\tau) = \tau(1-\tau), m(\tau) = (m(1) - m(0))\tau + m(0)$ and $T = 0.01$. The choice of a good initial approximation is the main matter when the parameters of the penalty function $\bar{\sigma}$ and ϵ become extremely small. In this case, it is helpful to apply a continuation strategy for the parameter ϵ, changing the value of this parameter from $\epsilon_0 = 10^{-1}$ to the desired value of ϵ. To highlight the advantages of this strategy, we solve the optimal control problem (15) choosing $\bar{\sigma} = 10^{-4}$ and two different values of $\epsilon = 10^{-3}, 10^{-6}$, the solution is plotted in Figure 5.

In Table 18 the results are computed without applying the continuation strategy, hence we observe that if on one hand only the codes bvp5c, tom and tomc fail for $\epsilon = 10^{-3}$, on the other all the codes do not converge for $\epsilon = 10^{-6}$. Consequently, in Table 19 we run the codes using the continuation strategy. All the numerical tests use an initial mesh of 16 equidistant points. For the continuation strategy in Table 19, except for acdc and acdcc, the parameter ϵ is initially set to $\epsilon_0 = 10^{-1}$ ($\epsilon_0 = 1$ for tom and tomc), and then it is changed using $N_\epsilon = 10$ logarithmically equispaced values up to reach the value required ϵ. However, to obtain the convergence of bvp4c for $\epsilon = 10^{-6}$, we put the value of $N_\epsilon = 100$ when $tol = 10^{-4}$ and $N_\epsilon = 20$ when $tol = 10^{-6}$ and for tom/tomc we put the value of $N_\epsilon = 55$. The conditioning parameters reported in Table 20 show that the problem is not stiff, but it is ill conditioned since $\kappa_1 > \kappa_2$.

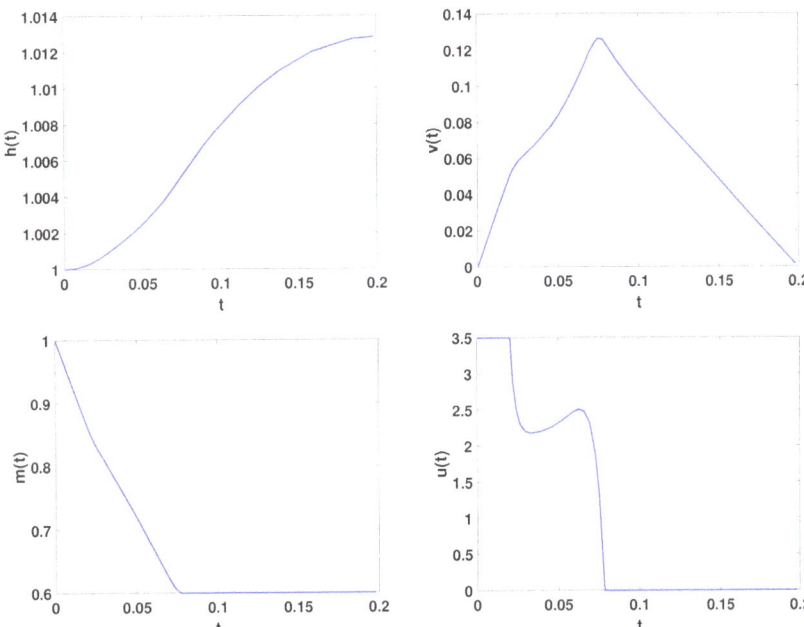

Figure 5. Goddard rocket, $\epsilon = 10^{-3}, \bar{\sigma} = 10^{-4}$: from left to right solutions in time for altitude h and mass m (on the **top**), for velocity v and thrust u (on the **bottom**).

Table 18. Goddard Rocket problem (15) solved using a piecewise quadratic penalty function with $\bar{\sigma} = 10^{-4}$ and $\epsilon = 10^{-3}$: final mesh (fM), total number of vectorized function evaluation (NVF) and mixed errors for h, v, m, T and u.

	fM	NVF	Error h	Error v	Error m	Error T	Error u
				$tol = 10^{-4}$			
bvp4c	1388	8058	8.9×10^{-10}	4.9×10^{-7}	4.7×10^{-9}	4.6×10^{-8}	3.0×10^{-5}
twpbvp_m	31	86	6.0×10^{-7}	3.8×10^{-5}	4.0×10^{-5}	1.1×10^{-5}	1.0×10^{-3}
twpbvp_l	31	91	1.0×10^{-6}	6.9×10^{-5}	7.5×10^{-5}	1.2×10^{-5}	1.7×10^{-3}
				$tol = 10^{-6}$			
bvp4c	1466	20,898	4.9×10^{-12}	2.7×10^{-9}	2.6×10^{-9}	3.1×10^{-10}	1.7×10^{-7}
twpbvp_m	32	142	4.6×10^{-9}	1.6×10^{-6}	1.5×10^{-6}	1.4×10^{-7}	9.8×10^{-5}
twpbvp_l	33	169	3.7×10^{-10}	6.8×10^{-8}	6.5×10^{-8}	3.3×10^{-9}	3.9×10^{-6}

Table 19. Goddard Rocket Problem (15) solved using a piecewise quadratic penalty function with $\bar{\sigma} = 10^{-4}$ and the continuation strategy: final mesh (fM), total number of vectorized function evaluation (NVF) and mixed errors for h, v, m, T and u. * Observe that acdc for $tol = 10^{-4}$, $\epsilon = 10^{-6}$ obtains a solution for $\epsilon = 3.98 \times 10^{-6}$.

				$\epsilon = 10^{-3}$								$\epsilon = 10^{-6}$			
	fM	NVF	Error h	Error v	Error m	Error T	Error u	fM	NVF	Error h	Error v	Error m	Error T	Error u
tol = 10^{-4}															
bvp4c	47	2823	3.2×10^{-9}	1.2×10^{-6}	1.1×10^{-6}	1.2×10^{-7}	7.9×10^{-5}	138	46,507	5.0×10^{-9}	3.8×10^{-7}	4.1×10^{-7}	4.4×10^{-8}	2.3×10^{-3}
twpbvp_m	16	268	7.0×10^{-7}	4.0×10^{-5}	4.4×10^{-5}	4.6×10^{-6}	8.5×10^{-4}	93	484	4.5×10^{-9}	3.4×10^{-5}	3.3×10^{-5}	2.4×10^{-7}	9.4×10^{-2}
twpbvp_l	16	304	1.1×10^{-6}	7.3×10^{-5}	7.9×10^{-5}	9.4×10^{-6}	1.7×10^{-3}	225	627	3.2×10^{-9}	5.2×10^{-7}	4.9×10^{-7}	5.6×10^{-8}	1.5×10^{-3}
tom	401	66	1.4×10^{-7}	6.8×10^{-5}	6.7×10^{-5}	1.7×10^{-7}	4.2×10^{-3}	541	224	1.6×10^{-8}	3.7×10^{-5}	3.2×10^{-5}	1.7×10^{-7}	4.2×10^{-2}
tomc	291	62	5.4×10^{-8}	3.9×10^{-5}	3.7×10^{-5}	8.7×10^{-7}	2.1×10^{-3}	286	210	1.1×10^{-8}	9.5×10^{-5}	9.2×10^{-5}	8.7×10^{-7}	1.2×10^{-1}
acdc	17	341	7.2×10^{-7}	8.2×10^{-5}	8.4×10^{-5}	9.3×10^{-6}	5.1×10^{-3}	24 *	2066	3.8×10^{-9}	8.7×10^{-7}	7.1×10^{-7}	9.2×10^{-8}	4.8×10^{-3}
tol = 10^{-6}															
bvp4c	148	9189	3.8×10^{-11}	7.7×10^{-9}	6.9×10^{-9}	9.1×10^{-11}	8.2×10^{-7}	1385	52,028	5.2×10^{-11}	2.9×10^{-9}	3.5×10^{-9}	7.9×10^{-10}	1.9×10^{-5}
twpbvp_m	29	593	2.3×10^{-8}	2.3×10^{-6}	2.3×10^{-6}	3.4×10^{-9}	1.4×10^{-4}	149	681	1.8×10^{-10}	2.9×10^{-7}	2.8×10^{-7}	8.0×10^{-9}	8.2×10^{-4}
twpbvp_l	32	646	4.5×10^{-9}	2.3×10^{-6}	2.2×10^{-6}	2.1×10^{-7}	1.4×10^{-4}	119	969	8.9×10^{-10}	2.9×10^{-6}	2.5×10^{-6}	1.1×10^{-8}	6.9×10^{-3}
tom	551	80	1.6×10^{-8}	1.4×10^{-5}	1.4×10^{-5}	3.0×10^{-8}	7.3×10^{-4}	661	279	6.3×10^{-9}	1.7×10^{-5}	1.5×10^{-5}	3.0×10^{-8}	3.5×10^{-2}
tomc	321	74	1.1×10^{-7}	1.7×10^{-5}	1.6×10^{-5}	2.4×10^{-8}	7.7×10^{-4}	731	286	7.3×10^{-10}	5.6×10^{-6}	5.4×10^{-6}	2.4×10^{-8}	1.2×10^{-2}
acdc	28	496	4.5×10^{-9}	2.3×10^{-6}	2.2×10^{-6}	2.1×10^{-7}	1.4×10^{-4}	58	942	2.3×10^{-10}	3.5×10^{-7}	3.4×10^{-7}	3.5×10^{-9}	9.7×10^{-4}

Table 20. Goddard Rocket problem: conditioning parameters computed using $tol = 10^{-6}$.

	σ	κ	κ_1	κ_2	γ_1
		$\epsilon = 10^{-3}$			
twpbvpc_m	4.8	9.1×10^2	7.3×10^2	1.8×10^2	2.2×10^2
twpbvpc_l	4.9	9.0×10^2	7.2×10^2	1.8×10^2	2.2×10^2
tomc	5.4	9.0×10^2	7.3×10^2	1.8×10^2	1.7×10^2
		$\epsilon = 10^{-6}$			
twpbvpc_m	5.3	1.0×10^3	8.2×10^2	2.0×10^2	2.0×10^2
twpbvpc_l	5.2	1.0×10^3	8.1×10^2	2.0×10^2	2.0×10^2
tomc	5.5	1.0×10^3	8.1×10^2	2.0×10^2	1.7×10^2

8. Minimization of the Fuel Cost in the Operation of a Train

As in [2,47] an optimal control problem in transportation is to minimize fuel cost in the operation of a train. To simplify the track is supposed to be straight. Let x be the position along the track measured from a fixed reference point and v the velocity of the train, such that the minimization problem is equivalent to solve the optimal control problem

$$\min_{v,u_a} \int_0^{4.8} u_a\, v\, dt,$$
$$x' = v,$$
$$v' = h(x) - F(v) + u_a - u_b, \tag{16}$$
$$0 \le u_a \le 10, \quad 0 \le u_b \le 2,$$
$$x(0) = v(0) = v(4.8) = 0, x(4.8) = 6,$$

where $F(v(t))$ models the friction due to the rolling of the wheels and the air resistance and $h(x)$ is the active component of the gravitational force due to hill slopes that are respectively defined as

$$h(x) = \frac{2}{\pi}\left(\tan^{-1}\left(\frac{x-2}{\delta}\right) + \tan^{-1}\left(\frac{x-4}{\delta}\right)\right), \quad \delta = 0.05,$$
$$F(v) = 0.3 + 0.14|v| + 0.16 v^2.$$

Moreover, the control variables u_a and u_b represent respectively the acceleration provided by the engine and the deceleration from applying the brakes.

First, as smoothing technique let us consider piecewise quadratic penalty functions defined as

$$P^a(u_a;\epsilon;\tau) = \frac{\epsilon}{2}(u_a - 5)^2 + \frac{1}{\tau^2}\begin{cases} (u_a - 10 + \tau)^2 & u_a > 10 - \tau \\ (\tau - u_a)^2 & u_a < \tau \\ 0 & \text{otherwise,} \end{cases}$$

$$P^b(u_b;\epsilon;\tau) = \frac{\epsilon}{2}(u_b - 1)^2 + \frac{1}{\tau^2}\begin{cases} (u_b - 2 + \tau)^2 & u_b > 2 - \tau \\ (\tau - u_b)^2 & u_b < \tau \\ 0 & \text{otherwise.} \end{cases}$$

so that the Problem (16) can be written as

$$\min_{v,u_a} \int_0^{4.8} \left(u_a v + P^a(u_a;\epsilon,\tau) + P^b(u_b;\epsilon,\tau)\right) dt,$$
$$x' = v, \tag{17}$$
$$v = h(x) - F(v) + u_a - u_b,$$
$$x(0) = v(0) = v(4.8) = 0,\ x(4.8) = 6.$$

From the Hamiltonian formulation we obtain the following BVP

$$\begin{aligned}
x' &= v, \\
v' &= h(x) - F(v) + u_a - u_b, \\
\lambda' &= -\mu h_x(x), \\
\mu' &= -\lambda + \mu F_v(v) - u_a, \\
x(0) &= v(0) = v(4.8) = 0, \ x(4.8) = 6,
\end{aligned} \quad (18)$$

where u_a and u_b, computed by solving the equations

$$P_{u_a}^a(u_a; \epsilon, \tau) + \mu + v = 0, \qquad P_{u_b}^a(u_b; \epsilon, \tau) - \mu = 0,$$

are respectively

$$u_a = \begin{cases} \dfrac{5\epsilon\tau^2 + 20 - 2\tau - \tau^2(\mu + v)}{\epsilon\tau^2 + 2} & u_a > 10 - \tau \\ \dfrac{\tau(5\epsilon\tau + 2 - \tau(\mu + v))}{\epsilon\tau^2 + 2} & u_a < \tau \\ \dfrac{5\epsilon - (\mu + v)}{\epsilon} & \text{otherwise,} \end{cases}$$

$$u_b = \begin{cases} \dfrac{\epsilon\tau^2 + 4 - 2\tau + \tau^2\mu}{\epsilon\tau^2 + 2} & u_b > 2 - \tau \\ \dfrac{\tau(\epsilon\tau + 2 + \tau\mu)}{\epsilon\tau^2 + 2} & u_b < \tau \\ \dfrac{\epsilon + \mu}{\epsilon} & \text{otherwise.} \end{cases}$$

As shown in Table 21, all the methods, starting with an initial mesh of 16 equidistant points and initial solution $x = v = \lambda = \mu = 1$, converge when $\epsilon = 1, 0.5$ and $tol = 10^{-4}$.

Table 21. Minimization of the fuel cost in the operation of a train (18) using a piecewise quadratic penalty function with $\tau = 10^{-2}$: final mesh (fM), total number of vectorized function evaluation (NVF) and mixed errors for x, v, u_a, u_b.

			$tol = 10^{-4}$								
			$\epsilon = 1$						$\epsilon = 0.5$		
	fM	NVF	Error x	Error v	Error u_a	Error u_b	fM	NVF	Error x	Error v	Error u_a	Error u_b
bvp4c	121	1747	3.0×10^{-6}	6.7×10^{-6}	3.4×10^{-5}	1.7×10^{-5}	116	2414	1.4×10^{-6}	1.8×10^{-6}	5.1×10^{-5}	6.3×10^{-5}
bvp5c	52	3259	9.8×10^{-7}	5.9×10^{-6}	3.3×10^{-5}	1.8×10^{-5}	56	4295	5.3×10^{-7}	4.8×10^{-6}	1.4×10^{-4}	1.5×10^{-4}
twpbvp_m	34	132	5.6×10^{-6}	2.3×10^{-5}	2.0×10^{-4}	9.1×10^{-5}	52	124	2.6×10^{-2}	3.3×10^{-2}	1.5×10^{-2}	6.1×10^{-3}
twpbvpc_m	47	132	5.7×10^{-6}	2.3×10^{-5}	2.0×10^{-4}	9.1×10^{-5}	55	104	2.6×10^{-2}	3.3×10^{-2}	1.5×10^{-2}	6.2×10^{-3}
twpbvp_l	33	136	9.0×10^{-6}	2.8×10^{-5}	4.3×10^{-4}	1.0×10^{-4}	223	124	2.2×10^{-2}	2.6×10^{-2}	9.9×10^{-3}	8.7×10^{-4}
twpbvpc_l	46	136	9.0×10^{-6}	2.8×10^{-5}	4.3×10^{-4}	1.0×10^{-4}	115	104	2.2×10^{-2}	2.6×10^{-2}	9.9×10^{-3}	8.7×10^{-4}
tom	1471	44	5.9×10^{-5}	3.9×10^{-4}	4.0×10^{-4}	1.4×10^{-4}	1091	45	4.9×10^{-6}	2.6×10^{-5}	1.1×10^{-4}	8.9×10^{-5}
tomc	1406	148	4.0×10^{-7}	1.5×10^{-6}	2.2×10^{-5}	8.7×10^{-6}	2896	93	1.2×10^{-7}	3.7×10^{-6}	3.5×10^{-5}	1.5×10^{-5}

Now, decreasing the value of ϵ, all these methods fail, since the Problem (18) is highly ill conditioned and strongly depends on perturbations. However, these methods can reach the convergence using a continuation strategy on the parameter ϵ. As initial ϵ we can choose 1 or 0.5, since we know that all the methods converge for those values. Moreover, we need to define the discretization for the perturbation parameter, namely we consider N_ϵ logarithmically equispaced points in the range ϵ_0, ϵ. Since the continuation depends on the choice of N_ϵ and the initial value ϵ_0, in Table 22 we show the results obtained using $\epsilon_0 = 1$ and $N_\epsilon = 5$, except for tom/tomc for which we need to consider for the convergence $N_\epsilon = 10$. Our interest is to analyze the performance of the codes for small perturbation parameters, as $\epsilon = 10^{-2}, 10^{-3}$, requiring an exit tolerance $tol = 10^{-3}$. In Figure 6 we show the solution for $\epsilon = 10^{-2}$. The conditioning parameters in Table 23 suggest that the problem is ill conditioned but not stiff, in fact $\kappa, \kappa_1, \kappa_2, \gamma_1$ are all much greater than 1. The condition number of the matrix of the last step of the integration procedure (last column of Table 23) is very high and confirms the ill-conditioning of the problem.

Table 22. Minimization of the fuel cost in the operation of a train (18) using a piecewise quadratic penalty function with $\tau = 10^{-2}$ and continuation strategy: final mesh (fM), total number of vectorized function evaluation (NVF) and mixed errors for x, v, u_a, u_b.

			$\epsilon = 10^{-2}$						$\epsilon = 10^{-3}$		
	fM	NVF	Error x	Error v	Error u_a	Error u_b	fM	NVF	Error x	Error v	Error u_a	Error u_b
bvp4c	69	6055	1.1×10^{-5}	3.4×10^{-5}	2.1×10^{-3}	1.9×10^{-2}	78	8213	1.3×10^{-5}	3.4×10^{-5}	5.2×10^{-3}	1.9×10^{-1}
bvp5c	39	7180	6.0×10^{-6}	6.0×10^{-5}	5.5×10^{-4}	1.6×10^{-2}	59	8012	1.3×10^{-6}	5.0×10^{-5}	6.2×10^{-5}	1.6×10^{-2}
twpbvp_m	415	432	4.0×10^{-6}	7.8×10^{-6}	1.2×10^{-3}	4.8×10^{-8}	589	319	4.1×10^{-6}	9.0×10^{-5}	2.0×10^{-3}	2.4×10^{-8}
twpbvpc_m	132	359	5.5×10^{-5}	3.8×10^{-4}	4.2×10^{-3}	2.4×10^{-2}	589	319	4.1×10^{-6}	9.0×10^{-5}	2.0×10^{-3}	2.4×10^{-8}
twpbvp_l	202	312	4.3×10^{-5}	3.9×10^{-4}	2.8×10^{-3}	9.0×10^{-3}	589	332	3.3×10^{-6}	8.1×10^{-5}	1.9×10^{-3}	1.6×10^{-8}
twpbvpc_l	202	312	4.3×10^{-5}	3.9×10^{-4}	2.8×10^{-3}	9.0×10^{-3}	589	332	3.3×10^{-6}	8.1×10^{-5}	1.9×10^{-3}	1.6×10^{-8}
tom	2201	99	1.5×10^{-6}	1.2×10^{-4}	1.7×10^{-4}	1.3×10^{-3}	2166	102	3.7×10^{-5}	6.8×10^{-4}	1.8×10^{-2}	1.4×10^{-1}
tomc	2211	218	4.8×10^{-6}	1.2×10^{-4}	1.1×10^{-3}	5.2×10^{-3}	2886	230	6.3×10^{-6}	5.9×10^{-5}	4.2×10^{-3}	2.2×10^{-1}
acdc	36	723	2.2×10^{-6}	1.2×10^{-5}	3.6×10^{-4}	9.4×10^{-3}	40	1219	9.4×10^{-7}	1.7×10^{-5}	7.6×10^{-4}	5.2×10^{-9}

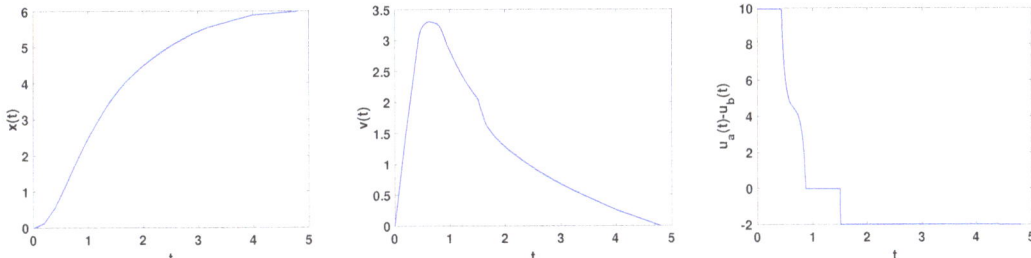

Figure 6. Minimization of the fuel cost in the operation of a train $\epsilon = 10^{-2}$: from left to right solutions in time for the position x, the velocity v and the difference between the control variables representing the acceleration and the deceleration $u_a - u_b$.

Table 23. Minimization of the fuel cost in the operation of a train: conditioning parameters computed using $tol = 10^{-6}$, cond is the condition number of the matrix associated with the last nonlinear iteration.

	σ	κ	κ_1	κ_2	γ_1	Cond
			$\epsilon = 10^{-3}$			
twpbvpc_m	6.6×10^5	9.9×10^{12}	9.9×10^{12}	1.1×10^3	1.5×10^7	9.9×10^{25}
twpbvpc_l	5.8×10^7	7.3×10^9	7.3×10^9	7.3×10^2	1.6×10^2	5.3×10^{19}
tomc	6.6×10^0	2.8×10^3	1.5×10^3	1.3×10^3	2.0×10^2	2.4×10^{15}
			$\epsilon = 10^{-6}$			
twpbvpc_m	1.1×10^8	5.0×10^{10}	5.0×10^{10}	4.7×10^2	4.9×10^2	2.5×10^{21}
twpbvpc_l	8.6×10^7	7.1×10^9	7.1×10^9	4.7×10^2	1.1×10^2	5.1×10^{19}
tomc	3.2×10^0	1.4×10^3	5.1×10^2	8.9×10^2	1.4×10^2	2.8×10^{14}

9. Conclusions

In this paper, after a review of general-purpose codes for solving boundary value problems we have solved some challenging optimal control problems derived using the indirect method. The presented results show that this approach could be a good alternative to the direct methods for the solution of this kind of problems, especially if the mesh selection strategy adopted is suitable for stiff problems in the case of hypersensitive problems, or an appropriate initial condition is computed for the nonlinear iteration using a continuation strategy. All these techniques can sometimes require the application of some regularization procedure, as in the presence of singular arc. Our goal with this paper is to give some indications useful to handle the input parameters of a BVP code to achieve an accurate solution, since the default values assigned usually works for very simple regular problems. Moreover, some codes give in output information about the stiffness and the conditioning of the problems, which could be used in choosing the correct solution method.

Author Contributions: Writing—original draft, F.M. and G.S. Both authors contributed equally to this work. All authors have read and agreed to the published version of the manuscript.

Funding: The research of Francesca Mazzia has been funded by the PON "Ricerca e Innovazione 2014-2020", project "RPASInAir: Integrazione dei Sistemi Aeromobili a Pilotaggio Remoto nello spazio aereo non segregato per servizi", n. ARS01_00820 and the research of Giuseppina Settanni by the INdAM-GNCS 2020 Research Project "Numerical algorithms in optimization, ODEs, and applications" (the authors are members of the INdAM Research group GNCS).

Institutional Review Board Statement: Not applicable.

Informed Consent Statement: Not applicable.

Conflicts of Interest: The authors declare no conflict of interest.

References

1. Bellman, R. *Dynamic Programming*; Princeton University Press: Princeton, NJ, USA, 1957.
2. Biral, F.; Bertolazzi, E.; Bosetti, P. Notes on Numerical Methods for Solving Optimal Control Problems, *IEEE J. Ind. Appl.* **2016**, *5*, 154–166. [CrossRef]
3. Rao, A. V. A Survey of Numerical Methods for Optimal Control (AAS 09-334). In *Astrodynamics 2009, Proceedings of the AAS/AIAA Astrodynamics Specialist Conference, Pittsburgh, PA, USA, 9–13 August 2009*; American Astronautical Society by Univelt: San Diego, CA, USA, 2010; pp. 497–528.
4. Soetaert K, Cash J, Mazzia F. *Solving Differential Equations in R*; Springer: Berlin/Heidelberg, Germany, 2012.
5. Ascher, U.; Christiansen, J.; Russell, R.D. Collocation software for boundary value odes. *ACM Trans. Math. Softw.* **1981**, *7*, 209–222. [CrossRef]
6. Ascher, U.M.; Mattheij, R.M.M.; Russell, R.D. *Numerical Solution of Boundary Value Problems for Ordinary Differential Equations*; Classics in Applied Mathematics Series; SIAM: Philadelphia, PA, USA, 1995; Volume 13.
7. Cash, J.R; Wright, M.H. A Deferred Correction Method for Nonlinear Two-Point Boundary Value Problems: Implementation and Numerical Evaluation. *SIAM J. Sci. Statist. Comput.* **1991**, *12*, 971–989. [CrossRef]
8. Bashir-Ali, Z.; Cash, J.R.; Silva, H.H.M. Lobatto deferred correction for stiff two-point boundary value problems. *Comput. Math. Appl.* **1998**, *36*, 59–69. [CrossRef]
9. Cash, J. R.; Moore, G.; Wright, R. An automatic continuation strategy for the solution of singularly perturbed nonlinear boundary value problems. *ACM Trans. Math. Softw.* **2001**, *27*, 245–266. [CrossRef]
10. Ascher, U.M.; Spiteri, R.J. Collocation software for boundary value differential-algebraic equations. *SIAM J. Sci. Comput.* **1994**, *15*, 938–952. [CrossRef]
11. Enright, W.H.; Muir, P.H. Runge-Kutta software with defect control for boundary value odes. *SIAM J. Sci. Comput.* **1996**, *17*, 479–497. [CrossRef]
12. Boisvert, J.J.; Muir, P.H.; Spiteri, R.J. A Runge-Kutta BVODE solver with global error and defect control. *ACM Trans. Math. Softw.* **2013**, *39*, 11. [CrossRef]
13. Boisvert, J.J.; Muir, P.H.; Spiteri, R.J. BVP_SOLVER-2. Available online: https://cs.stmarys.ca/~muir/BVP_SOLVER_Webpage.shtm (accessed on 25 May 2021).
14. Mazzia, F.; Cash, J.R. A Fortran test set for boundary value problem solvers. *AIP Conf. Proc.* **2015**, *1648*, 020009. [CrossRef]
15. Test Set for BVP Solver. Available online: https://archimede.dm.uniba.it/~bvpsolvers/testsetbvpsolvers/?page_id=27 (accessed on 26 February 2021)
16. Kierzenka, J.; Shampine, L.F. A BVP solver based on residual control and the MATLAB pse. *ACM Trans. Math. Softw.* **2001** *27*, 299–316. [CrossRef]
17. Kierzenka, J.; Shampine, L.F. A BVP solver that controls residual and error. *J. Numer. Anal. Ind. Appl. Math.* **2008** *3*, 27–41.
18. Cash, J.R.; Hollevoet, D.; Mazzia, F.; Nagy, A.M. Algorithm 927: The MATLAB Code bvptwp.m for the Numerical Solution of Two Point Boundary Value Problems. *ACM Trans. Math. Softw.* **2013**, *39*, 15. [CrossRef]
19. Mazzia, F.; Sestini, A.; Trigiante, D. The continous extension of the B-spline linear multistep metods for BVPs on non-uniform meshes. *Appl. Numer. Math.* **2009**, *59*, 723–738. [CrossRef]
20. Amodio, P.; Settanni, G. A finite differences MATLAB code for the numerical solution of second order singular perturbation problems. *J. Comput. Appl. Math.* **2012**, *236*, 3869–3879. [CrossRef]

21. Auzinger, W.; Fallahpour, M.; Koch, O.; Weinmüller, E.B. Implementation of a pathfollowing strategy with an automatic step-length control: New MATLAB package bvpsuite2.0. Tech. Rep. ASC 2019. Available online: https://www.asc.tuwien.ac.at/~ewa/software_development5.htm/ (accessed on 25 May 2021).
22. Auzinger, W.; Kneisl, G.; Koch, O.; Weinmüller, E.B. A Collocation Code for Boundary Value Problems in Ordinary Differential Equations. *Numer. Algorithm.* 2003, *33*, 27–39. [CrossRef]
23. Mazzia, F.; Cash, J.R.; Soetaert, K. Solving boundary value problems in the open source software R: package bvpSolve. *Opuscula Math.* 2014, *34*, 387–403. [CrossRef]
24. scipy.integrate.solve_bvp. Available online: https://docs.scipy.org/doc/scipy/reference/generated/scipy.integrate.solve_bvp.html (accessed on 25 May 2021)
25. De Marinis, A.; Iavernaro, F.; Mazzia F. A minimum-time obstacle-avoidance path planning algorithm for UAVs. *Numer. Algor.* 2021. [CrossRef]
26. Zong, L.; Luo, J.; Wang, M. Optimal Concurrent Control for Space Manipulators Rendezvous and Capturing Targets under Actuator Saturation. *IEEE Trans. Aerosp. Electron. Syst.* 2020, *56*, 4841–4855. [CrossRef]
27. Zong, L.; Emami, M.R. Concurrent base-arm control of space manipulators with optimal rendezvous trajectory. *Aerosp. Sci. Technol.* 2020, *100*, 105822. [CrossRef]
28. Putkaradze, V.; Rogers, S. Constraint Control of Nonholonomic Mechanical Systems. *J. Nonlinear Sci.* 2018, *28*, 193–234. [CrossRef]
29. Gerdts, M. *Optimal Control of ODEs and DAEs*; De Gruyter Textbook: Berlin, Germany, 2012.
30. Bader, G.; Ascher, U. A New Basis Implementation for a Mixed Order Boundary Value ODE Solver. *SIAM J. Sci. Stat. Comput.* 1987, *8*, 483–500. [CrossRef]
31. Capper, S.; Cash, J.; Mazzia, F. On the development of effective algorithms for the numerical solution of singularly perturbed two-point boundary value problems. *Int. J. Comput. Sci. Math.* 2007, *1*, 42–57. [CrossRef]
32. Cash, J.R., Mazzia, F. Efficient global methods for the numerical solution of nonlinear systems of two point boundary value problems In *Recent Advances in Computational and Applied Mathematics*; Simos, T., Ed.; Springer: Dordrecht, The Netherlands, 2011; pp. 23–39. [CrossRef]
33. Kierzenka, J. Tutorial on Solving BVPs with BVP4C. MATLAB Central File Exchange. Available online: https://www.mathworks.com/matlabcentral/fileexchange/3819-tutorial-on-solving-bvps-with-bvp4c) (accessed on 25 February 2021)
34. bvp5c. Help MATLAB. Available online: https://www.mathworks.com/help/matlab/ref/bvp5c.html (accessed on 25 February 2021).
35. Brugnano, L.; Trigiante, D. *Differential Problems by Multistep Initial and Boundary Value Methods*; Gordon and Breach Science Publishers: Amsterdam, The Netherland, 1998.
36. Mazzia, F.; Trigiante, D. A hybrid mesh selection strategy based on conditioning for boundary value ODE problems. *Numer. Algorithms* 2004, *36*, 169–187. [CrossRef]
37. Cash, J.R.; Mazzia, F. Conditioning and hybrid mesh selection algorithms for two-point boundary value problems *Scalable Comput.* 2009, *10*, 347–361.
38. Amodio, P.; Budd, C. J.; Koch, O.; Settanni, G.; Weinmüller, E.B. Asymptotical computations for a model of flow in saturated porous media. *Appl. Math. Comput.* 2014, *237*, 155–167.
39. Amodio, P.; Settanni, G. Variable-step finite difference schemes for the solution of Sturm-Liouville problems. *Commun. Nonlinear Sci.* 2015, *20*, 641–649. [CrossRef]
40. Amodio, P.; Settanni, G. Numerical Strategies for Solving Multiparameter Spectral Problems. In Proceedings of the Numerical Computations: Theory and Algorithms, NUMTA 2019, Crotone, Italy, 15–21 June 2019; Sergeyev, Y., Kvasov, D., Eds.; Lecture Notes in Computer Science; 2020; Volume 11974, pp. 298–305.
41. Pulverer, G.; Söderlind, G.; Weinmüller, E. Automatic grid control in adaptive BVP solvers. *Numer. Algorithm.* 2011, *56*, 61–92. [CrossRef]
42. Rao, A.V.; Mease, K.D. Eigenvector approximate dichotomic basis method for solving hyper-sensitive optimal control problems. *Optim. Control Appl. Methods* 2000, *21*, 1–19. [CrossRef]
43. Aykutlug, E.; Topcu, U.; Mease, K.D. Manifold-Following Approximate Solution of Completely Hypersensitive Optimal Control Problems. *J. Optim. Theory Appl.* 2016, *170*, 220–242. [CrossRef]
44. ICLOCS2 (Version 2.5). Imperial College London Optimal Control Software. Available online: http://www.ee.ic.ac.uk/ICLOCS/default.htm (accessed on 2 February 2021).
45. Bertolazzi, E.; Biral, F. Approximating Bang–Bang solutions in optimal control with indirect methods. In Proceedings of the Multibody Dynamics 2007, ECCOMAS Thematic Conference, Milan, Italy, 25–28 June 2007; Bottasso, C.L., Masarati, P., Trainelli, L., Eds.; Dipartimento di Ingegneria Aerospaziale, Politecnico: Milano, Italy, 2007.
46. Dolan, E.D.; Moré, J.J.; Munson, T.S. *Benchmarking Optimization Software with COPS 3.0*; Technical Memorandum ANL/MCS-TM-273; Argonne National Laboratory, Mathematica and Computer Science, Division: Lemont, IL, USA, 2004.
47. Vanderbei, R.J. Cases Studies in Trajectory Optimization: Trains, Planes, and other Pastimes. *Optim. Eng.* 2001, *2*, 215–243. [CrossRef]

MDPI
St. Alban-Anlage 66
4052 Basel
Switzerland
Tel. +41 61 683 77 34
Fax +41 61 302 89 18
www.mdpi.com

Mathematics Editorial Office
E-mail: mathematics@mdpi.com
www.mdpi.com/journal/mathematics

www.ingramcontent.com/pod-product-compliance
Lightning Source LLC
LaVergne TN
LVHW070612100526
838202LV00012B/627